Praise for

MW00983253

"*The Book of Noticing* is the p\ Hauswirth is the most precious walking companion, sharing moments of her life as a spouse, daughter, friend, mother and, especially, avid reader . . . this book reminds us that the first step in preserving the planet and our own well-being is simple observation. We must, as she does so well in these pristine essays, take notice."

—JILL SISSON QUINN, author of *Deranged: Finding a Sense of Place in the Landscape and in the Lifespan*

"In each chapter, Katherine takes us with her on her walks and brings us into her mind, her heart and her family as she explores the smallest of creatures, seeks answers to the simplest yet most complex questions, and shares both her own and her son's deep love of the natural world. This book comes at a time when humanity's relationship with the natural environment can either save or destroy our planet. It all begins with noticing."

—SUSAN RAUSCH, Owner & Director, Camp Earth Connection

"Katherine Hauswirth has come home from her walks through the woods with little drops of serenity and wonder. Now those collections, pooled together in this book, are deep enough to immerse readers."

—NATHANAEL JOHNSON, author of *Unseen City* and *Grist* staff writer

"Katherine Hauswirth's *Book of Noticing* rings and echoes with her love and longing for the natural world. Like the bobbing tentacles of slugs she describes, these insightful meditations vibrate at the frequency of change."

—ERIC D. LEHMAN, author of *Afoot in Connecticut*

"This book by powerful author Katherine Hauswirth is actually my religion. I don't mean that I just revere and respect the interdependent web of life she explores, I mean that I worship it. Her flowing words about thousands of threads in the tapestry of Mother Nature (that we might otherwise overlook or take for granted) are my idea of spiritual catharsis . . . ."

—CAROL HOLST, Founder of *Postconsumers.com*

"The strength of this collection of trail vignettes is Hauswirth's clear and friendly contemplative voice. I want to walk with her again soon, now that I have inherited her gift of noticing, as all readers who take up these pages will."

—AMY NAWROCKI, author of *Four Blue Eggs*

The
Book *of* Noticing

Collections and Connections on the Trail

For Change —
Wishing you
many happy moments
of noticing!

The
Book *of* Noticing

Collections and Connections on the Trail

Katherine Hauswirth

HOMEBOUND PUBLICATIONS
Ensuring the mainstream isn't the only stream

© 2017, Text by Katherine Hauswirth

All rights reserved. Except for brief quotations in critical articles or reviews, no part of this book may be reproduced without prior written permission from the publisher: Homebound Publications, Postal Box 1442, Pawcatuck, CT 06379.

WWW.HOMEBOUNDPUBLICATIONS.COM

Published in 2017 by Homebound Publications
Front Cover Image © Logobloom | Shutterstock.com
Cover and Interior Designed by Leslie M. Browning
ISBN 978-1-938846-73-1
First Edition Trade Paperback

Homebound Publications
Ensuring the mainstream isn't the only stream.

WWW.HOMEBOUNDPUBLICATIONS.COM

10 9 8 7 6 5 4 3 2 1

Homebound Publications is committed to ecological stewardship. We greatly value the natural environment and invests in environmental conservation. Our books are printed on paper with chain of custody certification from the Forest Stewardship Council, Sustainable Forestry Initiative, and the Program for the Endorsement of Forest Certification.

To Mommy, always supportive, always happy when I am.

To Tom, Gavin, and Linda, staunch, loving supporters who walk alongside and point the way to good things.

When one tugs at a single thing in nature, he finds it
attached to the rest of the world.
–JOHN MUIR

My work is loving the world.
Here the sunflowers, there the hummingbird—
equal seekers of sweetness.
Here the quickening yeast; there the blue plums.
Here the clam deep in the speckled sand.
–MARY OLIVER, in *Messenger*

Earth teach me quiet
as the grasses are still with new light.
Earth teach me suffering
as old stones suffer with memory.
Earth teach me humility
as blossoms are humble with beginning.
Earth teach me caring
as mothers nurture their young.
Earth teach me courage
as the tree that stands alone.
Earth teach me limitation
as the ant that crawls on the ground.
Earth teach me freedom
as the eagle that soars in the sky.
Earth teach me acceptance
as the leaves that die each fall.
Earth teach me renewal
as the seed that rises in the spring.
Earth teach me to forget myself
as melted snow forgets its life.
Earth teach me to remember kindness
as dry fields weep with rain.

–Earth, Teach Me (Ute Prayer)

Introduction

This morning I slept later than usual and wasn't out the door until 6:45. I have a vestigial tendency to wake up on farm time, perhaps a ripple pushing out from the side of the family that raised cattle in Virginia and, before that, staked a homestead claim and cowboyed their way around Wyoming.

No need for 4 AM rising here in Deep River, Connecticut, but still, now that summer days are upon us, my eyes usually pop open at about that time. By the time I splash my face, leash the dog, and replenish the knapsack with collapsible dog bowl, water, and my camera, there is just the faintest hint of burgeoning day visible through the window pane. With a jingle of Molly's collar we step out, past Oswald the rabbit's hutch and sounding a purposeful rhythm on the hollow, wooden porch stairs.

I slept later because it was a rare day off, and I knew I had nearly the whole 4th of July to get in my walk and maybe even try for a reprise in the cool of the evening. The neighboring village of Chester is setting up for "4 on the 4th," the annual 4-mile run, and I head towards the hilly back roads that will dump me out at the center of the town. I am not a runner, but the walk to Chester is curvy and lush and as an added prize I can gawk at the preparations underway.

Molly's curious and determined nose and her matronly, waddling weight keep the pace of our walks at the observant, unhurried end of the dial. We go for distance, not speed, and this means that my chronic awakening before dawn is a special gift on work days, at least the long days of the warmer seasons. I can immerse myself in a whole world with leisurely stride before my husband and son even open their eyes.

I am happiest when walking, but the happiness is not due to walking in and of itself. To be truly enjoyable my rambles have to be outside, relatively unhurried, and through a place that will give me some window into the goings on in nature. There's something about the feel of the air on my skin, about the smell of the honeysuckle as I walk past, about the rubber band chorus of frogs in the still pond that hasn't yet ceased to stir my soul. Each walk has its little surprises—a cluster of mushrooms that look like goo, a nearly completely intact oak insect gall, the surprising beauty of a humble brown slug against indigo asphalt, an inordinately leggy buttercup.

This is the kind of joy I want to hang on to. But I am a New Englander now, albeit a transplant from New York, and with the region comes some sense of tempering the abandon of summer with the reminder that the long, warm days are circling towards a cold, hard close. It's not just the northern climate that sobers me, though. I've got some life statistics that can't be ignored.

I'm now inarguably middle aged, and the tides of heat that sometimes flood over my body remind me that I'm entering another phase of life. I'm feeling panicky about passing the halfway point (and that's assuming I live until 90!). At the wise old age of 45 it dawns on me that, unless I attend to them, my twin passions for nature and words are at risk of gradually evaporating, like a glass of water left absently on a table in some largely unused room.

My husband Tom is just ahead of me on the path of life, several years older and starting the next curve. And far ahead of both of us is my mom, who still embodies a quiet beauty but can no longer converse with me or care for herself. Still fresh-faced on the journey is my son Gavin, approaching his twelfth year of life with ceaseless energy. His happy wandering is like Molly's mission of ambulatory zeal; it demands "look at this; smell this; taste this—you must not miss it!" This is a demand that calls to my inner workings. I'm determined to follow it.

John Muir said, "When one tugs at a single thing in nature, he finds it attached to the rest of the world." It's like that for me. My walks prove this to me time and again; they help me to connect, and not just with the proverbial great outdoors. Even when my mind seems to be on autopilot, I'll later find that some icon from a recent walk—a salamander, a hawk, the smell of green onions, holds more than just a pleasant image or lingering sense memory. If I take the time to pause and ponder, my walks are rife with totems that point me onto a more illuminated path, one where I am more aware of what matters most, in my own mind and soul but also in the larger world.

The walks also remind me that the more complex or troubling elements of my life exist in tandem with the steadfastness of the day. They are the yin to my yang of a fiery mix of sunset and the last October leaves, the rarely seen otter in Roger's Pond, even just simple warmth on my shoulder.

I've set up my own microcosm of the Museum of Natural History—well actually, I conceived of the idea but Gavin made it real, labeling a good-sized shoe box Cabinet of Curiosity. It sits inside our pantry window and is filled with mementos from many days outside—nests, shells, stones, seed pods. I've surprised myself, too, by arriving recently into the 21st century—my first smart phone has let me snap photos with one hand (since I hold the leash in the other) when I can't cram a find—like runny mushrooms and gorgeous slugs—into my pack.

I am not so hardy as to relish a walk in the dark and cold of winter. In fact, I've determined that my personal comfort threshold is at about 40° F (although I'm trying earnestly to reset it to 30° or below). Perhaps that's why I feel so driven to collect and catalogue mementos from my warmer weather strolls. I know that this winter, when the pavement is slick with ice, I'll be poring over my finds, the way gardeners leaf through the seed catalogue and fire up their basement grow lights. This *Book of Noticing* is a more permanent way of preserving my small discoveries, with the hope that by writing about their pull I may also share some of the truth and beauty that simple walks through a generous landscape continue to uncover.

Natural Neighborhood

Most of these adventures center on, or begin in,
Deep River, Connecticut.

———————————

The Ancients

When we were finalizing purchase details on our house, we stood on the back deck chatting with the sellers. I gestured toward the low stone wall at the back of the yard, and the forested hill beyond. "Whose property is that?" The husband shrugged, indifferent to the land—"Oh, that's state forest. The Cockaponset."

I still marvel at the fact that "State forest in your backyard!" did not make it into the real estate section's two inches of text. Instead, the advertisement used up word space with "Martha Stewart slept here." Well, she hadn't. And why employ false advertising when you have an amazing circuit of paths and hills beckoning you from a stone's throw? They could have had me at "state forest," despite the acrobatic troop of flying squirrels in our new home's attic.

Maybe I shouldn't judge the realtor's and sellers' nature attunement deficiency too harshly. We ourselves haven't taken full advantage of the Cockaponset's proximity. In fact, every time we are up in what Gavin nicknamed "The Top Woods," I spend a lot of time asking why we aren't back there every day. Maybe it's because my neighborhood walks remind me more of my childhood suburbia. Maybe, when the light is dim, I feel safer when more people are around, despite knowing that statistically I am much safer in the woods than on the asphalt. Maybe I'm avoiding the insatiable ticks that find me too frequently on the old logging trail back there.

Some days, I can walk through the forest and, while I feel a general appreciation for the variegated greenery and dappled sunlight, I just see trees—lots of them. I'm not fully there, not tuned into birds, bugs, plants, or any detail, really. I'm probably walking past myriad signs of wildlife that have walked the same path, maybe just minutes before.

Scratch "probably," for the constant presence of wildlife during even a brief jaunt into the woods is undeniably the case (assuming "wildlife" includes the very small). I so admire those naturalists who get down to the magnifying glass level. In *The Forest Unseen*, David George Haskell spent a year observing all manner of tiny life in a meter-wide mandala. A chapter in Alexandra Horowitz' *On Looking* is devoted to "Flipping Things Over," featuring field naturalist Charley Eiseman, a vigilant and enthused observer of insect (and other small creature) signs—tiny larval trails in a leaf, slug teeth marks, and such. This is the kind of guy who spends five hours in a driveway turning over leaves and logs before setting out on the "official" invertebrate tour he's planned.

The best naturalists seem to be steadfast and patient, with an admirable eye for detail. Serious birders, for example, are their own fixated subspecies, known for both precision and devotion. They watch and wait, sometimes only glimpsing a small portion of a wing in the brush. They take notes; they keep lists; they make sketches. They hold still for long intervals.

In the past, witnessing this dogged persistence, precision, and commitment has given me an inferiority complex. I start lists and lose them. I acquire shelf loads of nature books but forget the finer details of what I have learned, remembering instead the joy of the walk and then the eager paging through to learn, skimming text and pictures and relishing the promise of all that variety, that carefully composed knowledge. But I figured out more about who I am when I hovered over a sentence in *The World's Strongest Librarian.* Author Josh Hanagarne describes librarians: "As a breed, we're the ultimate generalists. I'll never know everything about anything, but I'll know something about almost everything."

This applies to my naturalist life: elm or sycamore? tortoise or turtle? heron or egret?—the details sometimes blur together in my less than scientifically rigorous approach to the species I encounter on my walks. But I've decided that being encyclopedic may be overrated. Being immersed in the moment of discovery and digging up a delicious fact or two in the infatuated aftermath may be, in fact, its own form of devotion, and just as worthy as the recounting of facts dutifully learned and diligently catalogued.

On one particular Sunday morning in the spring, I didn't have to be especially diligent to discover and observe life around me. It could be that the woodland creatures, like me, were relishing that particular day's moist climate. Maybe I was more alert that morning, or maybe it was the youthful senses of Gavin and his two fifth-grader friends, but the Cockaponset was thrumming with four-legged finds that Sunday. They ambled at us from all directions.

Most noticeable and abundant were the orange newts, which I have referred to for most of my life more generically as "salamanders." To be more technically precise, these were Eastern/red spotted newts, and they are the only newts found in Connecticut. They were in the "eft," or young and terrestrial, stage. Young efts, orange with distinctive, lighter spots and birthed in water, spend up to three years on land before returning to the pond for the adult phase of life. I was surprised to learn that these creatures can live for up to 15 years, and also that they secrete toxins, especially considering how many I had kissed gently on their tiny snouts during childhood summers in Vermont.

I have a treasure of an old book that's been keeping me company since early spring. *Sundial of the Seasons*, by Hal Borland, is a compilation of columns published in The New York Times from the mid-1940s through the early 1960s. I like to think about Hal, who for several decades lived on 100 acres in the Berkshire Hills of Connecticut, faithfully informing his readers about nature from the twilight of World War II through about the midpoint of the Kennedy administration. (And that's just this particular subset of columns. He posted columns for 37 years). His wise and appreciative commentary on the constancy of nature's cycles no doubt provided some comfort in otherwise tumultuous times. One review of a Borland book logged by nature writer John Hay summed up what appeals to me about the columns and their author. They embody a "spare and loving way of interpreting the daily miracles of the year, common mysteries faithfully recorded."

It sounds like Hal had many colorfully spotted walks like mine, for his May 19 entry in *Sundial* describes with familiarity a rainy May day lit by the "almost translucent" hues of earth-trodding efts. He describes the leisurely procession of these finger-length creatures, adding, with Biblical undertone, that "Biologically speaking, they are of the ancients, related to the amphibians, great and small, that crept from the waters of long ago and found the lands of this earth good."

I had a sense of the ancient when I watched the many newts that lined our path. The heavy-lidded, straight gazes out over noble noses recalled to me the quest of the dinosaur and the quietude of the Sphinx. It's unclear whether the eft phase works more to protect or extend the range of the species, but it's quite clear that the great mass exodus from and eventual return to the water is a pattern that's been observed for ages. This species possesses what's called "true navigation," although the exact mechanism of its internal GPS remains murky. Even when landmarks are removed, the eft can find its way back to the waters of its origin.

As I mused about the convergence of so much life on the same path where orange more commonly appears in the form of the odd mushroom or autumn oak leaves, our merry band of naturalists made its way to a minor, T-shaped intersection on the trail. The narrower road forming the stem of the T is rarely traveled, since its dead end abuts the shoulder of busy Route 9. But charging up the loamy road, front and center and in photogenic slo mo, was a fine, hardy example of the common snapping turtle. The ancient sensibility that had emanated from the past hour's orange legion paled in comparison to the mud-encrusted, gargoyle bulk of this iconic specimen.

Had we been a more educated bunch, we might have guessed that this was highly likely to be a female en route to (or from) an egg-laying venture, a mission in which she would dig a hole and deposit her clutch of up to 40 eggs. But I still have the video footage to prove how clueless we all were. It records us gleefully naming the snapper Phil, after considering Bob and Joe. The only smarts we had consisted of our hands-off policy with "Phil" the fertile female, fearing a sharp and lightning-fast removal of our digits. But we adored her from just a little way off, thrilled when she moved her flipper a bit, a sign of life to counter her stoic, stone-like affect in the presence of swarming gnats and humans.

That day in the woods left me with a heightened awareness of those small, persistent lives that mostly endeavor to hide from us when we venture up the hill. Sometimes we see the creatures face to face. On other occasions we derive their presence, like when we found both deer and coyote scat on another hike. These encounters, or near encounters, have brought me to my book shelf and computer keyboard with whetted curiosity, trolling pages and Web sites for both casual observations and scientific findings regarding the species freshest in my memory.

I regularly make another kind of delightful discovery when I stumble on like-minded photo snappers, bloggers, or species obsessives in my online research. At times when I take my morning walks, even through the more crowded and active sectors of Deep River, I see nary a soul walking, which makes me wonder if anyone ever lingers outside anymore. But the Hiker's Notebook postings, the i-naturalist cooperative nature logging site, virtual herbariums galore, even the official Connecticut Department of Energy and Environmental Protection site—these are proof for me that, like the often unobserved creatures going about their business alongside the relatively airless existence of many typical Americans, I, too, am part of an often unnoticed subtype: interested humans who have evolved to observe and record, and not primarily for the sake of objective science.

Rather, my naturalist compatriots and I aspire to reside in Hal Borland's honored camp, "interpreting the daily miracles of the year," faithfully recording with keyboard, pen, palette, and camera the common mysteries traversing the path just over the low stone wall.

Scent Trail

I've known this was coming. The start of the school year has come and gone, and the sun rises later in the mornings now. By December 21, daylight will have been whittled down to just over nine hours. Soon it will be the Southern Hemisphere's turn for longer days in the sun. My senses are restless.

My dog Molly's leash has a flashlight attached, and I've been mostly glad for it on our adventures in the purple-grey of predawn. When we approach a scrubby patch where I once happened upon a skunk, I flip on the light. When we near the bend where I recently noticed a dead baby possum, I click the button again to avoid treading on the poor hit-and-run victim. But I try to walk without the aid of electronic beam as much as I can, a bit vexed as dark fills more of my days by how very heavily I rely on my vision. I watch

Molly inhale the air, sifting it for some scent that completely eludes me, and I want to be more tuned in to that unseen, aromatic dimension. I'm trying to get my nose to step up.

Of course, an ocular-centric existence can provide unquestionable advantages. I've been thoroughly amused on the many occasions when I spot a rabbit and Molly, her beagle/fox hound mix making her especially "nosy," is consumed with purposeful joy at the scenting work before her, oblivious to the animal in the middle distance. Her entire body wags as she sniffs and chuffs. But she's still frantically tracking the tantalizing and tangled scent trail as the white tail hops away and under a shrub.

The nose is so much more powerful in most mammals than it is in humans, and recent research suggests that this is because humans don't form new neurons in the olfactory bulb following birth, maybe because there's been less and less dependence on our sense of smell over time. Dogs, however, are wholly "macroscented" —they have hundreds of millions of scent receptors, as well as an olfactory recess that can hold on to a smell longer and even a "backup" nose, called Jacobson's organ, above the roof of the mouth.

So, while Molly fails to sight the occasional wild rabbit, she's acutely tied to the data provided by her muzzle. This past summer she picked up on the scent of worms rent asunder by curbside mowers; they are a delicacy that she's learned to savor (and I have stopped gagging about it; we all have our predilections). She sorts the various molecules methodically with her nose and daintily picks a protein-packed snack out of the clumps of grass that spot our path. She also sniffs out a surprising array of decaying frogs, snakes, and turtles, considering how infrequently I spot these small corpses on our walks.

Only rarely do I pick up on particular smells during my rambles, and then I don't completely trust what the scent is telling me. From where does it emanate? Of what exactly is it reminiscent? All through the summer, around a particular bend, my nose picked up "cucumber." The scent, instantly recalling the slide of a rickety metal peeler down the length of a fresh pick and the reward of watery, seedy slices, was unmistakable. But I was stymied by the lack of any gardens down the whole length of that particular block. Then I learned that salad burnet, an herb of ancient origins, was used to line walkways during the 16th century so that the air would be perfumed with a distinct cucumber-like scent. It looks like a maidenhair fern; I thought I saw it tangled in the roadside brush.

What other remarkable plants have I missed because their particular scent didn't jump out quite as prominently? Only a handful of truly distinctive smells have captured my attention as I walk—warm asphalt in new rain, the candy-grape-bubblegum scent of flowering kudzu, stagnant muck at the edges of the marsh, sweet honeysuckle. I was surprised that my nose was astute enough to pick up on the fragrance of Concord grapes adjacent to the highway ramp not far from my house. I picked a sample and squeezed, watching the outer "slip skin" pull away to reveal the pale, inner fruit.

Here I indulged in an imaginative interlude. In 1966, my house was moved less than a half mile, up the road and around two corners to make room for the expansion of Route 9. Records don't show the exact location of the original site, but it's completely feasible that these grapes, growing on an unpopulated strip by the ramp overpass and just yards away from a stony little creek complete with miniature waterfall, are in the vicinity of my 1917 home's original dooryard. I imagined the first owners pulling a quick snack from the vine or carrying a bowl outside to gather the fruits for jam.

I thought, too, of Ephraim Wales Bull, who more than 50 years before my house was erected developed the Concord grape from wild ancestors that were particularly hardy (unlike their European counterparts) in the unforgiving New England soil and climate. His tombstone alludes to this claim to fame, which happened only after 22,000 seedlings were painstakingly cross-pollinated and rejected. But the epitaph also hints at the bitterness he reportedly felt about little recompense despite the public's enthusiastic embrace of his yield: "He sowed–others reaped."

Henry David Thoreau, along with his literary peers Ralph Waldo Emerson and the Alcott family, was a Concord contemporary of Bull and his experimental vines. This journal entry by Thoreau doesn't name the neighbor he describes, but he may have been referring to Bull: "Walking down the street in the evening I detect my neighbor's ripening grapes by the scent twenty rods off, though they are concealed behind his house." I wonder if Thoreau could have convinced Bull to view his lack of fitting compensation with less resentment, had he stopped by the arbors and shared his thoughts on another way of measuring accomplishment: "If the day and night are such that you greet them with joy, and life emits a fragrance like flowers and sweet-scented herbs, is more elastic, more starry, more immortal—that is your success. All Nature is your congratulation, and you have cause momentarily to bless yourself."

I draw strength from the words Thoreau so eloquently and earnestly penned. I'm sitting in a hotel room that looks out on a brick wall, but my mind is inhaling the roadside Concord grape and puzzling over Thoreau's word choice of "elastic" to describe a key characteristic of life. But then it occurs to me that by Thoreau's standard I am successful, that my life has within it a fragile but tangible elasticity.

Despite living a multitude of very mundane workdays, peppered with moments where I am too tired to pick up the house or face the task of paying bills, I also have many welcome stretches where I feel my spirit expanding, stretching to greet day and night with joy. For me some of the best moments, those with "fragrance like flowers and sweet-scented herbs," are quite literal. Yes, I rely quite heavily on my visual sense during my many ritual walks around Deep River. But when I attend to them, the scents that waft across my path bring me someplace deeper.

Science tells us that smell is the oldest sense, and I've learned that smell is unique in that the messages it conveys travel, post haste and quite deeply, into the brain. The olfactory bulb sits next to the hippocampus, the place where we create new memories of experience. The memories that scent triggers are instant and often leave us at a lack for broadly understood vocabulary. Just yesterday, I turned to my sister in a particular alcove of an antique shop and exclaimed, "This smells exactly like Poppy's garage!," a place that Linda knew had a lofty attic, an oversized wasp's nest, an abundance of bicycles, and a musty air of mystery and promise. But only three people on this earth have the identical frame of reference. "Poppy's garage" wouldn't mean the same thing to anyone else.

The science of the brain and the very personal and profound effect of smell translates into what will occur the next time I happen upon that roadside bunch of grapes, perhaps a full year from now when they are in a bright new season. I'll think of the first day I happened upon them, and my imaginings about my house before it was hauled off of its original foundation. I'll be reminded of Ephraim Bull's petulant and pithy epitaph, and the fact that there's a great Thoreau quote that forms a fitting caption for my grape-scented (and other aromatic) explorations.

Most likely, Molly will be straining at her harness, because the same wide cluster of thick brush that houses the grapes also conveys to her the promise of unseen creatures, maybe fox, maybe rabbit, certainly toads, worms, and snakes. Standing on the sidewalk and looking down into the cove of vine, leaf, loam, and running water, I'll learn from her deep immersion in reading the world with her nose, joining her in the primeval impulse to follow the scent wherever it leads.

Dead Reckoning

In July, I saw a dead rabbit under the pines flanking Bridge Street. I found no obvious external wound; the animal looked like it had just lain down for a nap. But rabbits don't play possum, of this I am sure. Something, be it age or infection or internal bleeding, had killed this fine, furry softness. It rested on a bed of long grass.

My walks took me by that same corner house for several weeks after, and I was rendered simultaneously uneasy and impressed by the transformation I witnessed there. The form became gaunt, just gradually, and I could imagine microbes processing its tissue from the inside. Maybe worms or maggots, too, although I was spared seeing them in writhing, multitudinous action. Eventually, a terrible, charry blackness overcame the form, and finally, only clean-looking bone remained, as if someone had assembled a skeleton from a model kit. A sensitive soul had placed a pine bough over the remains, and not long after that there was only a subtle dent, an oval in the grass where the rabbit had been.

The word that sprang to my mind when I reflected on this gradual dismantling of the rabbit was not the most obvious choice, on the surface. It was not *sad* or *scary* or *sobering*. The word was *efficient*, and I took some comfort in that efficiency. Although death is usually an unwelcome guest, it felt like by watching this return to the earth I was witnessing the order of things. No doubt a fresh new bunny was nearby, too, living its bouncy life, and this was an equally true and valid part of the equation. It is the way life on this earth works; the way it has always worked.

Any modicum of time spent in nature—or even driving past it—comes with reminders of death, although we seem to pay more attention and emote more when bigger creatures die. We feel pity and sadness for a deer or raccoon we see slain by traffic and now off on the side of the road, but we are conditioned to think little of the mosquito we swat, the tick we abolish, or the wayward moth or ladybug or fly corpses that we find on our windowsills. Sure, we might be squeamish about the cocooned meal in the spider's web or the maimed bug being carted away by an army of ants, but it isn't especially difficult to be philosophical about this, to frame it as a regrettable but unavoidable fact of life. The predators must do their preying, the insect kingdom must continue its expected cycle of generation and eradication, and on it goes.

Being this matter-of-fact about our own demise isn't quite as easy. Even in the face of fervent spiritual beliefs, death does a number on the survivors in its wake. Loss makes us wonder about existence, and what it means, and whether we will ever stop feeling blindsided and bereft. We think about our own eventual ends with anxiety. Somehow, it feels like we should be exempt from the brutality, the finality of loss.

But is it final? Professor Eric D. Lehman, in a reflection on his visit to Walt Whitman's grave, remembers how the poet's famous volume, *Leaves of Grass*, "refers to that 'green hair of graves' that covers the American landscape in a ubiquitous sheet. Each blade is distinct and individual, as we fancy ourselves to be. But underneath the surface, the rhizomatic roots are all connected." Lehman quotes Whitman's "Song of Myself": "You will hardly know who I am or what I mean,/ But I shall be good health to you nevertheless,/ And filter and fibre your blood." The professor recounts his experience of standing in Camden, reflecting on the words and the grave before him: "...I tried to think of Whitman meaning this literally, picturing the decaying nutrients of his body fostering the grass, the trees, the heron, and myself. This transformation seems obvious and basic, but we as a species have always shied away from it."

Whitman's way was one of overflowing enthusiasm, even about death and decay. No one can shout it from the rooftops quite like him, but the more I read about nature the more I come across observations in the same vein. Fast forwarding from Whitman's 1892 grave to present day, Professor of Biology Ursula Goodenough writes about death in a much more scientific manner, about how some simple cells—organisms like the amoeba or tumor cell, which don't have death programmed into their systems, are referred to in scientific parlance as *immortal*. She adds that a life cycle like the human one, with a germ line—a genome that gets passed on to the next generation, is a different story, in which "immortality is handed over to the germ line." After explaining the science, she muses, "Death is the price paid to have trees and clams and birds and grasshoppers, and death is the price paid to have human consciousness, to be aware of all that shimmering awareness and all that love. My somatic life is the wondrous gift wrought by my forthcoming death."

Just yesterday, one of the warmest spring days we've had so far, Gavin and I walked a couple of blocks from his art lesson on Main Street to Fountain Hill Cemetery, a sprawling, hilly resting place where famous, infamous, and ordinary people's lives are marked with cool granite and carved words. Most infamous among them is XYZ, a notorious and still anonymous bank robber who was gunned down in our town in 1899. We have a famous artist, too—Sol LeWitt—world-renowned for conceptual art and locally lauded for designing a striking synagogue in Chester, the next town over. Mostly, the markers commemorate people we don't know at all. Some spout poetry or are captions to elaborate statues; many simply bear a name and a date.

Of course, some of the stones demand an especially pensive pause—markers for young veterans who only got to come home in a casket, or the large family who perished together in a fire back in 1938. One recent addition of a large, white statue depicts a kneeling angel weeping, her face in her hands, her flowers dropped before her. She marks two deaths that happened on the same day, a man and a woman. These carefully etched markers remind me that death is the real deal. There is no way around the sorrow it brings.

Fountain Hill is also overflowing with life. Its manmade pond is a favorite spot for school children, who peer down into the murky water looking for tadpoles and water snakes. In the right season frogs are abundant, and with the right timing you can watch a great blue heron taking off, zooming over to nearby Pratt Cove. When the leaves are down and you walk back toward the Jewish cemetery, you can see the river below. Yesterday I saw a muskrat dive down and duck under the edge of the pond—presumably entering his unseen den.

Over the years, we have relished sightings of foxes and coyotes here, too. We have heard large acorns hitting the pavement, touched the soft bark of the cedars, and leaned down to watch slugs feeding atop large mushroom caps. We have wondered what creatures have dug precise holes in the moss and when we will again be mesmerized by a sinuous snake breeding ball in the pond. All of this among thousands of symbols of loss, a microcosm of the lively gifts and wrenching grief that visit every life. This duality resides within a couplet from a Ute prayer, which asks the Earth for lessons in dealing with both pain and joy:

Earth teach me acceptance
as the leaves that die each fall.
Earth teach me renewal
as the seed that rises in the spring.

When I walk Molly around Fountain Hill, taking care to let her leash out in the green spaces not yet occupied with memorials, I am learning as I go. Snatches of the words from Ursula Goodenough come to mind: *trees and clams and birds and grasshoppers... shimmering awareness...somatic life...forthcoming death.* I take instruction from those who went before, and in equal measure from the thriving life that presses on, touched by sun and wind and birth and flight. Here among the gray stones peppered with lichen, there is so much living to do.

The Wordless Writer
Baby Steps in Mindfulness

Yesterday I dropped Gavin at his art lesson and used my free hour to revel in the early morning sunshine. But at first my walk was a traveling inner monologue: "Do I have enough cash to pay the art teacher and buy an iced coffee? Where did I put my keys? Is my phone charged? Oh yeah, I have to reschedule that appointment…"

Many of my walks start this way, with noisy inner ramblings, until I start to become more permeable to my immediate surroundings. I feel the cool breeze on my arms. I hear a woodpecker high above and wonder where it is, exactly, and what its incessant drilling pursues. I take in the old stone wall and the fuzzy catkins on the pussy willow, idly searching the treetops for the hawk's nest I've been hearing about.

When the chatter in my head (affectionately known as "monkey mind" in several meditation traditions) finally dies down to a dull roar, a job hazard pops up. Words that I might want to write make their debut, soon followed by edits to said words, and then a list of magazines that might like them.

I learned that after the Buddha famously attained enlightenment while sitting under the Bodhi tree, he engaged in walking meditation. Words from Thich Nhat Hanh, a modern-day Buddhist, capture for me the impulse to walk toward a place of greater peace: "Do you know how many dirt lanes there are, lined with bamboo, or winding around scented rice fields? Do you know how many forest paths there are, paved with colorful leaves, offering cool and shade? They are all available to us, yet we cannot enjoy them because our hearts are not trouble-free, and our steps are not at ease." Lately I have been experimenting with walking mindfully, taking my own shaky steps toward ease in an amateur nod to the Buddhist tradition.

It can be surprisingly challenging to find that ease, even with the noblest of intentions. So I feel compelled to share a little trick that helps me. I know it sounds odd, but, mentally, I don what's basically an imaginary astronaut suit and pretend I have been sent on a "mission" to acutely observe. I have to be fully here and just observing; I can't be distracted or make to-do lists because I am committed to the mission and must take in every detail. I think the space suit part of the scenario is my mind's way of insulating the acute focus on the moment, of protecting this fragile and fleeting state of hyper-awareness.

I turn up my noticing so that I am more conscious of the rise and fall of my breaths; sometimes I even perceive them as having that whooshy, intensified space suit sound.

I continue to walk—breath in, breath out, perceiving myself as having that deliberate, slo-mo astronaut gait, head turning from side to side to take it all in. (I hope I don't unconsciously adopt Neil Armstrong's exaggerated, springy moon cadence as I walk meditatively in my invisible space-suit. That would be embarrassing!).

For a few minutes in my "one giant leap for mankind" mode I get to be in the moment, appreciating being and not planning any doing. I have noticed that my "doing" mode is often accompanied by a barrage of words in my brain—words that label, words that analyze and categorize, words that criticize and question and assume and debate. When I use my space suit trick my brain is quieted for a time, and in that quiet space I can see and hear and feel much more clearly.

The muting abilities of my imagined costume made more sense when I listened to an interview with Jill Bolte Taylor, a neuroanatomist who authored an account of having a stroke that basically wiped out the words in her mind. In *My Stroke of Insight,* she describes the experi-ence as if someone pressed the "mute" button on a remote control. During the first crucial days of recovery following brain surgery, she remembered nothing, and there were no words expressed, no words understood. But Taylor recalls the experience of being without words as one of joy. She describes simply experiencing the present moment, sensory intake, and feeling connected with everything around her. I get to have fleeting moments like this during my walking meditation. Yes, these moments are quite transient, but they feel like time in another dimension, a place of great peace and possibility.

The last time I had such a moment, the landscape widened and I noticed a particular spot where the sunlight was glinting magnificently, and how the leaves did a little shimmy in the wind. My focus zoomed in more precisely on a verdigris cluster of lichen on a tree stump, and then on tiny, perfect plant life sprouting stealthily amid grains of curbside sand.

It's hard for a writer to admit that wordlessness can be a good thing, but here I am, sharing my contradiction and my spacesuit. Maybe someday I won't need my Apollo 11 trick anymore. I'll simply step outside and take it all in, eyes and ears open, mind quiet, feet moving in thoughtful cadence across the welcoming curve of the Earth.

The Giving Tree

There's a small triangle of woods along busy Bridge Street, around the corner from my house. It's messy and untended, a neglected space between the Route 9 overpass bridge and a neighbor's yard. Molly and I walk by it on our many strolls, and since last winter the space has been bisected by a large fallen ash tree, its wide mass of roots tilted up into the air.

Poring over the roots up close, I see that remnants of shredded, windblown paper and plastic bags cling to the edges. Despite this bargain-basement decor, a robin has started her nest in the recess below the upper rim of the base. I'm taken with the stones that the roots grew around, embedded like gems in the filigreed disc that's come up into the light.

I Googled "things embedded in tree roots" and was disappointed that I didn't find a lengthy list of treasures—I imagined precious remnants of prior decades and centuries secreted in the circuitous tangle of loam and wood. It's much easier to find examples of how tree *trunks* grew around things—one Web site displays photos of a bicycle, a bench, a license plate, and strands of Christmas lights, to name just a handful of the myriad incidents in which trees have slowly enveloped inanimate objects.

Of course, to find out what buried roots may be hiding, you have to wait for the tree to topple over with such force that even its deepest, most tenacious grasp on the earth surrenders. There is one stunning photo, "Spiritual Roots," featured on the *National Geographic* website. It's a Buddha statue head from an abandoned temple in Thailand amid a tightly woven root system. The tree took the Buddha completely in—an apropos joining considering his famous moment of enlightenment while sitting in the shade of a fig tree.

I've been a Northeasterner all my life, and I am struck by what seems to me an oddity: people from Western desert or plains terrain can feel claustrophobic when they move to densely wooded areas, homesick with longing for the unbroken line of sight to the horizon. I also read that in some areas there are so few trees that they are known on an individual basis by their human neighbors. This is one of many other scenarios where things that are scarce attain more recognition and value—"dime a dozen" occurrences engender only a yawn.

Doubtless I have yawned my way past many trees. But, maybe due to more encounters of late, individual trees seem to be calling for my attention now. The alligator-skin–like bark of the fallen ash had me picking through a thorny thicket for a closer look. I examined the leaf litter and small sprouts starting there, and also the ropy vines braiding all around the area, wondering about the circumstances of the ash's uprooting and how long it would be before the trunk merged back into the earth.

There is another large, downed tree just up the hill in the forest, and I have an innate impulse to touch, and even nestle into, the broad, placental plane at its base. I love the smell of the humus that clings to beards of long, thin roots, thirsty probes that once reached far below the surface to seek moisture. I also delight in the textures of barks and in finding the denizens they harbor—spiders and fungi and a multitude of wriggling new lives, some of whom tunnel hieroglyphic-like paths as they travel the trunk. I've started to look more closely at the shapes of leaves and their endless variegation of color, and not just during the autumn foliage extravaganza.

I examine the galls that trees drop, especially the large, fat oak galls so prominent in my neighborhood. Galls: new tree tissue that grows in response to chemical signals and creates a cozy house for some embryonic insect—are probably less familiar to most of us than are acorns or seed pods, and the variety of gall shapes, sizes, and residents can fill a book (and in fact have filled several). My personal favorite is the wooly oak gall—small fuzzy spheres that bring me back to my childhood in New York suburbia, where they fell in the hundreds onto our concrete driveway. I never saw, and wasn't even remotely aware of, the tiny wasp larvae living inside.

J.R.R. Tolkien created characters called ents in his *Lord of the Rings* series—tree-like creatures who guard the forest. They can walk; they can attack and defend. But, outside of Middle Earth, trees cannot step away or fight back. The growth of galls, the enveloping of park benches and Buddhas, the stretching of roots, the dropping of acorns—all can be explained by the inability of plants to pick up and go.

I admire the quiet relentlessness of plant life—yes, even those invasive species described as overtaking and bullying the rest of the flora. And I treasure the way that plants, especially trees, continue to give so actively to the earth—both before and after their demise. I marvel at the gradual total disintegration helped along by mushrooms, termites, and other creatures until, eventually, even the lightest touch to the rotted trunk has it returning in soft, silent, sodden chunks to the ground.

We learn in grade school about the oxygen cycle, and about the lumber that makes our houses, furniture, composition books, fences, and cereal boxes. But most of us think less often about how plants protect and nurture us. They improve air quality, stabilize land, and clean toxic soils. They yield medicines, and perhaps even being in their presence can heal us. In Japan, there's a popular practice called "forest bathing"—basically immersing oneself in forest surroundings. Studies suggest that this practice can boost cancer-fighting immunity.

Gavin has a children's book called *Cactus Hotel* that gets at the exhaustive list of gifts from a plant. It's about a tree-*like* plant growing far from Connecticut—the iconic saguaro cactus, which grows Arizona's state flower and whose many arms are instantly recognizable to even non-desert dwellers. For many years, the cactus "hotel" bears fruit, feeding bats and birds. A jackrabbit gnaws on the cactus' pulp. A Gila woodpecker drills a hole and lives in the plant; an elf owl lives in another hole. And on it goes; more arms grow and hold a bounty of "hotel guests." Two centuries later, the dead cactus is downed by a strong wind. But it continues to support many comers and goers—lizard, termites, a scorpion. Collectively it nourishes or shelters an impressive segment of the desert ecosystem. It's hard not to feel stirrings of gratitude and affection for such an endlessly magnanimous plant.

What I am feeling toward trees and their cousins is what E.O. Wilson termed biophilia—"love for all living things"—and I see no reason why it should be confined to the human experience. Take Molly, for example. When we walk her eyes grow wide and interested when we encounter another life—fellow canine, human, feline, chipmunk, even a larger bug—and she seems to also relish her contact as she sniffs a low pine or strolls, snuffing, beneath the leggy wildflowers, through high grass. Sure, hounds follow their noses and an ancient hunting instinct asserts itself through these behaviors. But to my eye it's about much more than finding food. Molly's tail stands up high and wags when she's had a "meet and greet" with some other being. She's interested, more alive after the encounter. I feel the same way when I get to connect with other forms of life.

What if we take this a step further? For a while now I've entertained the fantasy that trees are thrilled when they have some incidental contact with me, in the same way that I am thrilled when I happen upon a deer on the path or mourning doves sunning on the driveway gravel. Often the bough of one of our large pine trees brushes the top of my head when I walk by, and I can imagine the tree taking its own moment to admire me—maybe my mobility, my noises, or my apparent tendency to muse—and being glad for the connection in passing. I continue to enjoy this thought despite gentle ribbing from my family. I thought of a comeback to their chuckles: Olympian snowboarder Jamie Anderson hugs trees before her runs, to help "transfer energy from the earth" to herself and soothe her nerves. (I'm not sure whether she thinks the trees get anything out of it.)

Just as some may assume that biophilia is a human-owned thing, I also think it's assumed that those creatures who cannot speak must be mindless or emotionally limited. But when we get to know our wordless pets we start to venture some intuitive guesses at an inner life, and I don't think our conclusions are pure fantasy. There is an active life of the mind underway, and palpable emotion among our dogs and cats and parrots. Why can it not be the same for trees and other plants?

I'm fully aware that there is no nervous system per se within a plant. But is it possible that there is another way to feel, to have an inner life, than just the brain and spinal cord? Is it hubris to assume that animal biology is the only acceptable format for experiencing meaning? Daniel Chamovitz, the author of *What a Plant Knows,* points out that plants see color, sense weight, and can smell disease on a neighboring plant, albeit with sensors and systems that seem foreign to us. What else might the plant be processing within its network of leaves, stalk, and roots?

Another kids' book with a plant as the star, *The Giving Tree*, has always left me with mixed emotions. You could interpret it as another story in praise of the benevolent role I admire for the saguaro cactus, or really the generous life cycle of nearly any botanical specimen. But in this story, the Giving Tree gradually gives up nearly all of its parts for the boy it loves, talking about how it's happy to help the boy. To me, the gradual hacking away, the selfless diminishment to a stump reads like a metaphor for an unhealthy relationship, like a spouse who enables a chronic, abusive partner or a parent who never voices an emphatic "NO," thus raising an entitled tyrant of a kid.

"Really?," I think, as the Giving Tree happily gives away its limbs, and then its trunk. "You're just here to serve, serve, serve until you are a flat, splintered whisper of your former self? How about asking for something in return? How about speaking up for yourself, showing the boy how to grow into a more thoughtful man?" But of course, in the real world, trees cannot talk, even to tell us they are absolutely willing to give it all. And we all know that in many places, home and abroad, the trees *have* given it all. We've already lost about two thirds of the rainforests—the source of so many species, livelihoods, and potentially many medicines unique to those environs, too. Our "intelligent" species has also destroyed many of the forests here in the United States, although, thankfully, reforestation has been on the rise.

Hugging trees may remain largely the purview of famous snowboarders and aging hippies. But recent reports show that some significant change has started, with more to be done. An article in *The New York Times* reflects on improved efforts by Costa Rican and Brazilian authorities to protect the countries' rainforests. Here at home, scientists are at work resurrecting the chestnut tree, which was driven to extinction by blight, and The Nature Conservancy has launched the Plant a Billion Trees campaign in response to rampant global deforestation.

I hope we can continue to find ways to return the many offerings the trees have bestowed, harvesting them thoughtfully and protecting their budding new generations. Whether motivated by love for the trees or the drive to better preserve our environment and our future—and perhaps it can be some of both—I would find that a much more satisfying ending to the Giving Tree story.

What Lies Between

There's a particularly well-manicured piece of property at the corner of Union and Village Streets. I don't know its resident very well. I know she hosts one of the prime stops for trick-or-treaters every year, opening her porch so ghouls and fairy princesses can select from among several generous candy prizes. I also know that she employs a local landscaping service to keep her shrubs pruned and her grass lush and verdant. The finery of her plot, topped off with whimsical statues, stands out among the other squares of land, which are generally well tended and adorned here and there with red, yellow, and purple flowers, but not nearly as picture-perfect.

When I walked by Mrs. Halloween's yard yesterday, a small, paperback-book-sized swath of curly river birch bark stood out against the freshly-mowed, deep green turf. The dingy white bark with salmon-colored underside looked out of place, crying out for attention in the otherwise manicured surroundings. So I relieved the yard of its small, gnarly raft, appropriating what the tree had discarded.

My find wasn't simply the size of a small book—it was something I could read. I could leaf through its pages, which had some approximation of a spine adhering them at the center. Between the papery leaves I found a waif of a greenish spider, a postage stamp-sized stretch of web, and a cluster of what looked like dry, husky seeds. I released the spider, who looked ready for a fight, back onto the lawn.

I looked up the "seeds" and learned that they were the first "skins" left behind by a troop of tiny molting baby caterpillars. Browsing further, I found photographs of gorgeous trails reminiscent of Sanskrit, tunneled into existence by larvae on birch trees. Dedicated gardeners are unlikely to view these devoted carvers of serifed script primarily as talented calligraphers—they present a true threat to the carefully tended arbor. But my research got me wondering about how much is going on between things. What am I not seeing as I walk, because it's found a snug home between protective layers?

One habit that may someday get me into trouble is peering into hollow logs, containers tailor-made for snakes wanting a cozy bed or perhaps an incubator for their eggs. But there's something irresistible about a window into a tree, once it has fallen. And how little I know, despite my tendency for intense and habitual peering! I was recently amazed to look up into a still-standing, hollow tree and discover that the branches crisscrossed all the way through. This didn't match the many cross-sectioned diagrams I've seen of tree trunks—bark, cambium, sapwood, the inner heart of hard-wood, solid and countable rings—a layer cake of birthdays. Did this particular tree's branches extend extra enthusiastically—and in retrograde fashion—once the cellulose had rotted away? Do squirrel residents enjoy leaping from rung to rung inside the secreted corridor, vaulting up to the attic to hide fat new acorn acquisitions? I am guessing my tree with branches inside was some kind of yew—this type of tree is unique in that it becomes hollow over time and can then extend new shoots down through the hollow, eventually growing what is, in effect, a tree within a tree.

And then there are the destinations that nestle within other, more mundane-seeming places, a multitude of secluded gardens to be found by the stalwart wayfarer. Tucked away behind an unassuming hilly neighborhood in my town is a sewage treatment plant, and a friend told me that there's a great shortcut to the river behind it. When I followed her directions, I suddenly found myself in a green space akin to the emerald runways between hedgerows that abound in Great Britain, a nation marked by centuries of boundary and thoroughfare. In his book *The Old Ways,* Robert Macfarlane recounts a winter walk along a favorite path: "there's a feeling of secrecy to it that I appreciate, hedged in as it is on both sides, and running discreetly as it does between field and road." I too, was feeling that I had my own secluded hedgerow, albeit with the rare concrete mole hill marked "sewer."

The greenery on either side of the corridor was thick, and I enjoyed the illusion of strolling down a long, private hallway of turf and tangle, glimpsing the river below through occasional, unleafed windows between branches. Gradually

I came out onto the railroad tracks, a former bustling span where now only the tourist steam trains run from time to time. Farther on, after Molly had ducked down under many brambles to lustily pursue fresh rabbit scents, we veered left off the tracks to the town dock, where I cooled off from my happily enclosed adventure and looked out over the water.

I thought about one of my early writing assignments when I first moved to Connecticut. I was to write about Manchester's Oakland Trail, and what I finally got published was a mildly self-deprecating riff about the ambitious ramble of a fledgling writer—clumsy, new to the state, and flummoxed by the prospect of taking notes while avoiding foot hazards of unfamiliar terrain. Despite the bead of nervous sweat that formed on my young cub reporter lip as I crossed crumbling industrial asphalt and climbed down onto a brookside trail, I was able to relish the hidden way. At first I found mostly graffiti and winced at the sound of roaring eighteen wheelers over a concrete bridge. But before long the path opened up and showed me its comelier curves. I had that same satisfaction I was to have later when I thumbed through the book of wisdom written in river birch dialect, and again when the treatment plant served as my gateway to adventure. I found the particular root of the satisfaction difficult to encapsulate, and then I remembered another real-life experience that gets at it fairly well.

There's a place called Nature's Art not very far from here, popular with kids for the gargantuan dinosaur that graces the parking lot, its vast collection ranging from real gems to cheap souvenir trinkets, and opportunities to pan for "gold." Downstairs they have a diamond saw, and for the right price you can pick a geode from among many overflowing, dusty pails and have them slice your prize open.

I've seen this play out on several occasions, and at one very precise moment I envy the diamond-saw operator his job: he cups the sliced halves of the rock between his pressed together palms, careful not to spread the seam. Before the reveal he looks with great seriousness at his customer, very often a wide-eyed child, and intones, with endearing and sincere excitement, "You are the first person in the whole history of the world, the first person EVER, to see inside of this geode." And then he opens up his hands to show a trove of crystals.

I know that I am not a frontier adventurer, that I am trodding land that lost its virginity long ago. No new maps are sprouting from my footsteps. But I hope that I always find ways to see, with alert and appreciative eyes, the places that might be otherwise go unnoticed. What lies between is a worthy destination, not any less so because it can be reached easily, hiding in plain sight alongside our pruned landscapes and sun-bleached pavements. It is ready to astound us, opening both hands to us in quiet revelation, like the childhood church that waits for release from our folded palms.

Pining

Pine (verb): to yearn intensely and persistently especially for something unattainable
Pine (noun): any of a genus (*Pinus* of the family Pinaceae, the pine family) of coniferous evergreen trees that have slender elongated needles and include some valuable timber trees and ornamentals
—Merriam-Webster Dictionary

I have a small stack of books parked on the corner of the dining room table, for ready reference as I plan my day at breakfast time. Among them are two treasured volumes, *Twelve Moons of the Year*, by Hal Borland, and *A Walk Through the Year*, by Edwin Way Teale, that I found side by side at a garage sale. I treasure them especially on days when work and school schedules or saturating rain showers conspire to prevent me from spending any significant time outside. Their value soars on the coldest and snowiest winter days.

Each book has dated daily entries reflecting on nature in my home state of Connecticut—Borland's based on his farm in Sharon, in the lower Berkshires (northwest in the state), and Teale's a love letter to Trail Wood, his property in Hampton (northeast). The two properties lie about 90 miles apart. I wonder if Borland, who lived in The Nutmeg State from 1945 until his death in 1978, knew Teale, whose *A Walk Through the Year* was published the year Borland died. They were born one year apart, both in springtime, the senior Teale in the very last year of the 19th century.

As the end of spring drew near, I wanted to see what these thoughtful authors might have to say about it. The *Twelve Moons* entry for June 20 is titled in its author's straightforward columnist voice: Summer Solstice, referring to what is coming the next day. Then he gets more poetic: "Ripening berries are sweet to the tongue. Brooks are languid, rivers are leisurely, bogs teem with brief amphibian and insect life." Teale's entries have no title. His prose for the same date is circuitous, meandering through "the faint, faraway jingle of bobolinks" and a rapturous impression of "tranquil skies, tranquil earth, tranquil everything" before circling back to acknowledge, wistfully, that a new season would greet the next day.

In each solstice eve entry I read an acute awareness of fleeting time intertwined with a sense of longing, and this resonates deeply. Teale summons a scene from Margaret Craven's *I Heard the Owl Call My Name*, describing how a main character walks riverside and repeats, wanting to hang on to the perfect spring day, "Don't go—not yet—not yet." Borland captures the inevitable pendulum of seasonal time: "the sun has now achieved its greatest northing and now turns back…Time has no resting place…Summer is briefly ours…"

Each man is taken with the lush beauty around him, but each is also voicing, with a touch of resignation, the inability to preserve the delectable spring in its full splendor, despite noble attempts with artful prose. The soulful subtext of their words brings me back to summer nights on the beaches of Long Island when I was young, climbing unmanned lifeguard stands with friends and gazing out past the cooling sand to the Atlantic, a wordless longing filling my silent throat. The salt-scented air, the dampness on my skin, the ceiling of stars, the expansive silence—it was rich and perfect, and it was coming to a close. School days waited just around the corner, but the ache I felt wasn't about the reprise of academic expectations. I wanted to stay in that moment of deep connectedness, to spend hours on that worn and splintered stand watching the frothy ocean spilling onto the shining beach. I wanted to forever forestall the jump down into the shifting hillocks of sand and the loose, barefoot stumble back to the manmade planes of shoes, pavement, car, and highway.

Long Island is now a distant shore across the Sound, and it's been decades since I commandeered an abandoned lifeguard perch. But summers remain a vehicle of longing for me. The longing is born from the ingrained habit of an anticipatory life, from acute awareness of the fleeting quality of these wondrous and luxurious long days. When I can manage to live in the moment and not pine my days away, there is a lovely feeling that comes from rising pre-dawn and hurrying out the door so I can walk into the sunrise—seizing a generous slice of time before I even have to entertain the prospect of myriad daily obligations. If only every day could start with a miles-long walk in warm and welcoming air!

My longing for nature is never completely sated, even if its fever pitch wanes a bit when the temperature turns chilly. Of late I've started to wonder if the amped up impulse to cling, to linger is a sign that I am attuned with what Gaian teacher and Buddhist scholar Joanna Macy calls "The Great Turning"—shorthand for our current age, one that's suspended between a society shaped by industrial growth and the possibility of a new one that is life-sustaining. Macy transmits hope for our ailing world in many ways, but she wonders aloud about the direction in which we will collectively turn. I, too, am unsure, but I seek out comfort in gestures of both adoration and action performed by those who, like Macy and me, are smitten with love for the world.

Along with my expanding naturalist sensibility comes a strong urge to examine things up close and catalogue what I see in detail. With this mission in mind my thoughts go immediately to the world of botany, because I tend to have trouble with the details of plants—the general impression is that most are green, some are big, and some are tiny. I can pick out only a dismally limited number of specific specimens.

Last week I took a walk here in Deep River, past great swaths of brush and woody knolls, and I mused that I wanted to view plants the way I view words—for a writer that means with great appreciation but also great precision. In the world of words, it matters greatly if you say *advocate* or *helper* or *believer*, even if all three words are listed in thesauri everywhere as synonyms. Why does it matter? Because everything changes according to the choices we make for our stories. Everything can change again with the stroke of the editor's pen.

In the plant world, we can literally read the leaves of botanical change and sometimes prevent negative outcomes, or even facilitate a positive shift in nature's balance. For example, one of the easier invasive species to identify is purple loosestrife, a pretty perennial with a ferocious tendency to take over the shores of native wetlands. The National Wildlife Refuge Association lists the species threatened or already affected by this invader, including 44 kinds of native plants that serve as higher-quality nutrition for animals, some federally endangered orchids and swamp rose mallow, the bog turtle, the black tern, and the canvasback duck. What to do about this problem, even if we can't reverse the encroachment and can only prevent complete takeover, is guided in this case by national agencies. Of course, what to do isn't always so straightforward and prescribed when it comes to ecology. It's a careful balance and it's not always clear how much human intervention is advisable, or if we might inadvertently encourage another problem. But knowing, without doubt, what the purple loosestrife looks like is an excellent place to start.

Successful ecological actions are often underpinned by that stewardship impulse that comes with loving the world, but protection is not my only motivation for cataloguing. As I begin to look more closely at the variegations of green and the shapes of the leaves around me, I revel in the learning and the exploring that my new quest will bring. I am excited by the prospect of getting to know intimately fellow creatures I could only describe with the broadest of adjectives before. *In Central Park in the Dark*, Marie Winn writes about the progression from beginning to experienced birder: "For the most part, [the beginners] notice only the gross differences...: big versus small, black versus red, a big fat bill...versus a sharp little bill...When I first started birding, I couldn't tell the difference between a white-breasted nuthatch and a black-capped chickadee...Now they look as unalike to me as an elephant and a giraffe."

For a while now, I've had the impulse to catalogue pine trees in the neighborhood. I think this arose from a visit to our local Christmas tree farm, where someone has tacked boughs from all of the Christmas tree varieties to a board, captioning them with black magic marker labels: balsam fir, Fraser fir, blue spruce, Douglas Fir, white pine. Plus, the block I live on and the adjacent road are resplendent with what I have been calling tall "pines," and even my rudimentary observations, casual looks while walking by, suggested to me great variety. At first I entertained the idea of setting out with scissors and snipping my own collection of boughs, until it occurred to me that I could simply snap some close-ups with my phone, avoiding harm to the trees and odd looks from the neighbors.

Reflecting back on that informative Christmas tree farm sign, I realized that I was using the term "pine" the way an amateur birder uses the term "sparrow," when in fact there are at least 33 sparrow varieties in the United States alone. I learned about a term that even experienced birders use, either dismissively or wistfully (as, in they sincerely long to figure out exactly what they are seeing) when they see some sparrow or warbler, wren or finch that defies precise description: "a little brown job. "

When considering my piney surroundings, which at first seemed to consist of all "big green jobs," I realized that there were spruces and firs to consider, too, an acknowledgment of my inaccurate habit of calling basically every needled tree a pine. Spruces, firs, and pines are all conifers. All produce cones (so they are not all, technically, *pine* cones!). But pine tree needles are attached in clusters of 2 or more, while with spruces and firs the needles are attached individually—not too dissimilar from how people hang tinsel on their Christmas trees, in clumps, or neatly strand by strand.

Here is what my investigation yielded: a Northern white cedar, a baby Eastern larch (aka tamarack), an Eastern hemlock, a Norway spruce, and a juniper—and this doesn't include at least three conifers that remained nameless because of my blurry photos. My detective work was greatly aided by a biology student's online field guide to conifers, which included a decision tree—needles or scales? clusters or singles? number of leaves in the cluster?, etc. All the words for what I once thought were "pines" got a bit confusing but I gave myself partial credit, because the taxonomic division is Pinophyta and the class is Pinopsida. Cedars and yews are the only varieties within this class that are not in the same family as pine, spruce, fir, larch, and hemlock.

Along the way I took a side trip to learn about the spittle bugs adorning the lower branches of my "Christmas trees." I was astounded to learn that there are more than 23,000 varieties of these mostly unseen creatures, who as nymphs secrete a frothy liquid disguise that also keeps them hydrated. They hatch from eggs that overwintered in dead twigs or bark, and they only emerge from the froth when they are full-fledged adults.

Returning to my botanical study, I learned that white cedars are also called arborvitae (tree of life) because they were known to cure sailors' scurvy. *Hackmatack* is another word for tamarack, and it is an Abenaki word for "wood used for snowshoes." My research confirmed my observation that evergreen conifers are a favorite shelter for all kinds of birds, especially in wintertime, since they can provide good cover while the deciduous trees are bare. Often I hear the loudest and most concentrated dawn chorus emanating from the Norway spruce right across from my front door.

Somehow, I had never made the connection that many birds, as well as squirrels, eat seeds from conifer cones. Studies have proven coevolution at work between the birds and squirrels and their cone breakfasts—the cones have adapted defenses over time so that in areas with mainly bird predators, the cones have thicker scales, while in areas dominated by squirrels the cones are heavier. So far there's no proof that squirrels have evolved to better conquer their cone "prey," but the research shows geographic differences between crossbills, according to the local pine cone population: birds in areas with thicker pine cones have deeper, less curved bills.

I've wanted to learn more about Deep River's ubiquitous conifers for years, so it's strange timing that as I write this I'm receiving a salvo of letters from The Nature Conservancy with this subject line: "Red Spruce Forests Need You, Katherine!" In Central Appalachia, almost a century after "The Great Cut"—clear-cutting that left only 5% of the red spruce forests intact—the Conservancy has set a goal of restoring the spruces so that creatures including saw-whet owls, West Virginia Northern flying squirrels, and Cheat Mountain salamanders have a safe and life-sustaining habitat. If I were to hop onto the Appalachian Trail here in Connecticut and foot it down to Maryland or West Virginia, I'd walk right into the landscape that's sparked this cause for concern.

All of my "pining" gives me a new sympathy for a relatively remote swath of our nation bereft of its red spruce. I am taken with the steadfastness and variety of the conifers in my area, and happy to think of them sheltering a host of creatures, many of whom I may never spot behind the thick branches. I have good frames of reference for the creatures the Central Appalachian red spruces host, too. I assume the species must have subtle, regional differences, but I can overlook that to recall the vastly adorable but overly-fertile flying squirrels that were our upstairs attic neighbors for years, and the legions of salamanders that carried on their orange parades in many forests that I have loved. Ever since reading *Central Park in the Dark* I feel an intense need to find a sleeping saw-whet owl with my binoculars, a challenge because they are painted with perfect camouflage. I know they are there—I have heard them!

I've relished my micro-expedition into the pines (okay, conifers—it's going to take some time to warm to the more technically correct but less audibly pleasing term). Another explorer, Sebastian Copeland, whose expeditions have been anything *but* micro, said "we will not save what we do not love," and that rings true when I think about the conifers around my neighborhood and the drastically thinned red spruces farther south. For years I have felt an instinctive pull toward the prolific, fragrant needles of evergreens, but this instinct seems to have deepened into love, a love that grew because I stopped to look at them, really look at them quite closely, and to learn more about their quiet, purposeful lives.

The Summer Solstice, that time of heightened light and longing, will be long past, coming into autumn when the pollinated female cones of the Norway spruce in my yard become fully mature, ready to spread their progeny. This is the first year that I will be watching intently for them. There's a bench just below, and I'm looking forward to sitting, watching, and waiting, leaning back and gazing into the tangle of brown and green, wondering where and when they might take root.

Sprung

My mother dusted off one of her favorite phrases with the start of every March: "spring has sprung." I like the cheeriness of this pithy pronouncement, and the suddenness, the burst open floodgates of the new season that it celebrates.

If any season seems simultaneously long in arriving and quite sudden, it is spring. We want it, we want it more, we agonize over its elusiveness and then it seems one day it is just here. Whether or not our calendars back us up, we feel it, hear it, can practically taste the pollen, and our hearts are often warmed long before we are officially in the new season. The sun on our skin, the birds working on their songs, flowers tumbling over themselves to blossom—these perennial blessings have a way of sneaking up. The only analogy I can think of is from the winter season. When I was a kid, I wondered if Christmas would ever really arrive.

Why was December 25th so far away? Then, the dazzling morning struck like a falling star—Santa Claus had finally appeared in all of his red velvet glory. But those mornings of glittering surprise pale next to the lingering, effervescent mood that burgeoning spring can create.

Of course, science tells us that there is so much incremental preparation going on behind the pre-spring scene. Reading nature journals penned by several writers, I am impressed by how many comment on the first hint of hope carried by lengthening hours of light, just a little more each day. The light brings a palpable sense of new possibility, even on frigid mornings. It's as if the stage lights are gradually warming in anticipation of a gala performance. The "stage lights" are our sun, and as the Northern hemisphere turns toward it the hibernators are yawning, stretching, and suddenly very hungry. The plants are channeling their energy into born-again experiences everywhere. The performers are warming up, setting the stage for a show-stopping finale rife with cascades of chlorophyll and varied progeny.

In some cases, though, it's like the performer just couldn't wait and ran pell mell onto the stage while the orchestra was still tuning up. Take mushrooms. You walk the dog one moist and breezy morning and the lawn is free of them. The next morning, they are covering the front yard. (Of course, mushrooms aren't only a springtime occurrence. But we notice them more as the weather warms and our time outside expands).

According to the *Handy Biology Answer Book*, mushroom's fruiting bodies (just one fraction of the fungi—the most visible part that appears above ground) are encouraged by warm, damp weather, and initially tiny round caps can indeed expand voluminously overnight. The book cites the extreme example of the stinkhorn *Dictyophora indusiata* or *Phallus indusiatus* (it does indeed look like a phallus, with a fishnet skirt on)—which pushes out of the ground at about 0.2 inches per minute. If you happen to be there when this happens (and that would mean you were someplace tropical), you might actually hear a crackling sound that comes from the fungus' rapid expansion outward and upward.

Vernal pools are another crescendo of spring sprung. The pools don't appear nearly as rapidly as *Phallus indusiatus,* but the transformation from fallow-seeming depression on the land to full-blown creature nursery tiptoes up to surprise us with an embarrassment of abundance. The Vernal Pool Association, in Peabody, Massachusetts, likes to refer to these annual miracles as Wicked Big Puddles, "wicked" being New England speak for "awesome." I wish I had coined this pithy slogan for vernal pools; it's quite satisfying to say it aloud, ideally while crossing planks on a boggy trail that leads to a wicked soup of teeming life.

Obligate species that vernal pools host (species that must breed in the pools) include the spotted salamander, the wood frog, and the Eastern spadefoot toad. There are plenty of facultative species, too, like the green frog, the Northern spring peeper, and the Eastern American toad. Facultative means that the species might breed in a vernal pool, but they might also choose another aquatic habitat.

I hastened to the Environmental Protection Agency Web site on March 1st, by then as desperate for growth and greenery as a shipwrecked castaway for fresh water. My thirst for busy new life was temporarily quenched by a heartening image of a blue, green, yellow, and white scene—a ring of bright blooms—and the text told me something I hadn't realized: "In the spring, wildflowers often bloom in brilliant circles of color that follow the receding shoreline of the pools" (at least in the type that inhabit grasslands). The pools themselves seem such a cup-running-over gift, and the floral "gift wrap" that lingers long after seems such an artful, and impossibly generous, bonus.

I made it my mission to find my own vernal pool this spring, one that I could visit at intervals, watching life unfold. I consulted Ann Courcy, a new friend and proud blogger at Barking Frog Farm. It turned out she works at SchoolMates preschool, a nature-oriented center on the grounds of nearby, beloved Bushy Hill, a 700-acre nature center that hosts a wide variety of camps and other programs.

On our first walk together at Bushy Hill, Ann led me to a rather narrow and not especially stunning drive-side pond—not a vernal pool but a modest body of water with moss on its banks and two impressive clusters of frog eggs—I could see the tiny black dot of a tadpole-to-be within each one. Then we departed off down the Blue Trail, which led to a bona fide vernal pool complete with instructional sign. Not only did it promise salamanders, grey tree frogs, spring peepers, and American toads, but it explained that even in dry summer months this depression shows traces of the richer, springtime wetland in the form of nearby blueberries and skunk cabbage.

Ann reported wood frogs calling at this pool not long ago, but on the day we visited the shallow circle seemed to be basically a big, silent, leaf-filled puddle. I took Gavin back a few days later—same thing. Of course, visiting these places during the day is not always ideal. Although they are harder to see at night, at least you can count on the frogs and toads calling out their presence after sunset.

Eager for the audio, we took a family trip over to Woodbury, a Connecticut town about an hour northwest. The Flanders Nature Center there was hosting a vernal pool talk and exploration one rainy April night, and we had our flashlights at the ready. The pool got quieter as we approached, the peeps slowing to an occasional, high-pitched comment. Poking our way around branches and brambles, we peered down at wood frog egg clusters. The kids who came equipped with muck boots (as well as Gavin, who didn't) walked slowly and gingerly between tiny mud and grass hillocks, as well as in the water, carting to shore a delicate spring peeper, a comparatively giant green frog, and—the *pièce de résistance*—a pair of wood frogs (seemingly oblivious to the crowd) copulating. Well, technically, they were in *amplexus*. The male stays on the female's back for up to 72 hours. He releases his sperm into the pool water as the female ovulates, and—*voila!*—fertilized egg mass. I guess it's worth the 72-hour "bed-in," as up to 2000 eggs are formed with each session. That's what I call a productive use of time!

Over the course of the pool tour I developed a massive desire to acquire muck boots of my very own, to maximize the delight of stirring the mud and discovering life. Although I was bootless and squished about in my soggy sneakers, I had finally acquired something else that I had coveted, and the next day I relished a long look at it in the daylight: *A Field Guide to the Animals of Vernal Pools* is produced by the clever Wicked Big Puddle promoters along with other Massachusetts organizations. The photos are a delight, and thanks to one in particular I now count a yearning to see the pearl-like eggs of the blue-spotted salamander as a topper on my wish list. The text confirmed for me, yet again, the marvelous unseen, the busy life thriving under cover of leaf and log: "A moderate sized vernal pool might have several thousand wood frogs entering to breed and then returning to the forest. Yet most people, even those who spend extensive time in the woods, never encounter one of these woodland creatures…salamanders are seldom observed except on rainy migration nights when hundreds might be moving to or from a vernal pool. Yet these animals live out their 20 years of life within a few hundred feet of that pool."

With all of this pop-up life inhabiting my synapses, the meandering trail that is my mind had me recalling a class I took on the Old Testament in college. I've retained very little of the content, but one tidbit that stuck was the professor's sharing of a theory I'd never heard before. She suggested that the manna from heaven that appeared suddenly one morning to the fleeing Israelites in the desert could actually have been bird droppings. I have long since forgotten which type of bird was supposed to have such palatable droppings, and my attempt to flesh out that story has yielded little scientific support. But I did find a treatise by Steve Kubby on the Internet about how manna may have been wild mushrooms, and the author put quite a lot of thought into how the fungi match up with the highly specific Biblical descriptions. He wrote: "For one thing magic mushrooms are small and round, and since they sprout so rapidly they would seem to appear overnight, as if out of the sky. Also, anyone harvesting them would immediately notice that they turn blue where torn and had no roots, giving more reason to believe that the mushrooms were of celestial origin. Note that manna does not just fall from heaven, but instead it is described as coming with the frost and dew, during the wet seasons." Another theory about manna published by *Smithsonian* magazine likens it to a sweet-tasting secretion of a variety of plant lice that infects certain shrubs in the Sinai Desert, and this theory seems to have legs (so to speak) since Bedouins on the Sinai Peninsula continue to harvest and eat the stuff. I can imagine the Israelites' surprise the first time they awoke to find the mysterious manna, whatever its origins. The Biblical description is for me a metaphor for the sudden-seeming new life that visits every spring with such abundance, heralded by a swelling of moisture (as in melting snow and April showers): "The manna came down on the camp with the dew during the night." (Numbers 11:9, NLT).

Miraculous as it was, it turns out the people weren't all that thrilled with the manna. They craved meat, and this part of the Bible describes them as a downright whiny crowd. Case in point, Numbers 11:5-7 (NIV): "But now we have lost our appetite; we never see anything but this manna!" This part of the story brings to mind something I had been noticing in myself. I love my time outside so very much. I am happiest when given unfettered time to wander, to observe, to muse. But sometimes I am just itchy for a sighting, one of those lovely photo-op moments when a creature crosses my path or an exotic-looking bloom shouts out at me in full color. Manna, though miraculous, isn't good enough. Bring on the meaty moments!

In the woods and pools of April, life is less obvious than I might choose. I often end up looking skyward, turning my head towards the birdsong, because the forest and the water seem largely a brown, orange, and green mosaic that keeps mostly mum about the life it nurtures. On a recent walk through a cedar swamp at Bushy Hill, I was thrilled to see just a small cluster of new frog eggs after tramping about for quite some time. Doubtless there were plenty of other hidden eggs, as well as creature mothers, fathers, siblings, cousins, predators, and predatees, but spotting them would require tons of time and patience. I never seem to have both in stock at once, and wonder whether the instant gratifications of the world beyond the woods have primed me to expect an on-demand sighting of whatever is on my wish list that day.

Words from Barbara Hurd, in *Stirring the Mud*—a whole book of essays on mucking about in swamps and bogs— remind me of nature's wisdom in being invisible often, even if it means I have to be content with imagining the creatures to whom April gives rise, with occasional sightings if I am lucky. Hurd's words about her favored watery haunts can apply equally to the much more ephemeral vernal pool:

> "To love a swamp, however, is to love what is muted and marginal, what exists in the shadows, what shoulders its way out of mud and scurries along the damp edges of what is most commonly praised. And sometimes its invisibility is a blessing. Swamps and bogs are places of transition and wild growth, breeding grounds, experimental labs where organisms and ideas have the luxury of being out of the spotlight, where the imagination can mutate and mate, send tendrils into and out of the water. "

This is an oft-quoted phrase from Hurd's book, and I noticed that the word "blessing" is at its center. Again, my mind veers from trail to words, thinking about a recent book my husband brought home. In A *Blessing of Toads* Susan Lovejoy writes about the many small creatures of her garden, and her title invokes a strong and immediate sense memory of what has felt like a blessing on many occasions: holding a cool, pulsing, soft-bellied, warty American Toad between two hands, where it has sat at my mercy, completely noiseless.

I know from my recent class with FrogWatch USA that the American Toad has an impressively lengthy, high-pitched, vibrating trill. I know that I should get my hands dirty and then hold him by the hips next time, so he doesn't have to absorb any lingering chemicals or get injured when he tries to launch away from my grasp. And I know that he will sing when he is ready, and not a moment sooner. Like spring and its myriad revelations, it will be worth my wait.

At Home and Work

When I can't get out on one of my treasured walks, sometimes just a few steps beyond my door—or even around the parking lot at work—have let me learn from nature.

Slug Fest

My first memory of slugs is a memory of murder. I couldn't have been more than eight when my neighbor friend Sara, even younger, informed me that slugs die when you coat them with salt. If I participated in the slaughter I have no memory of my guilt. I like to think I put up a vigorous protest. But I also remember being morbidly fascinated by the instant desiccation.

Slugs don't have an easy way in the world, at least not in the suburbs. They eat vegetables and ornamental plants alike and can creep up to the bases of these prizes of the back-yard patch, making them a sworn enemy of the gardener. Go ahead, Google "how to get rid of slugs" and astound yourself with the myriad assassination plots that arise on your search page. Drown the slug in beer, bloat it to death with cornmeal, sanitize it with ammonia, adding insult to squeaky-clean injury as it takes its sad, final breath.

A slug crossed my path just the other day, and I was moved by a certain earnest innocence expressed by its wobbling stalks, eyes bobbing atop twin tentacles. It looked determined—perhaps that determination was the will to get away from my looming presence, or a vow to survive despite centuries of persecution. Its presence moved me enough to have me extracting my phone from my pack to snap a photograph. I call it "Brown Slug on Dark Blue Pavement" and like to study the pattern on the glinting skin that reminds me of the whorls in my thumbprint.

The essence I derived from some reading up on slug science is that slugs are of the earth, quite literally. They are immersed in the dirt; they marinate in and draw sustenance from the juices surrounding them. And here is where I find something to envy. Many times I wish I were so permeable as to take the earth into me — memorize and somehow recon-stitute, wear on my skin, the dusky odor of the leaf pile; the specific, spongy texture of moss; the suction-cupped crawl of the impossibly ornate fuzzy caterpillar. Of course, being permeable is a double—and acutely—edged sword. Take in the delectable wonders of the world and you may also take in the salt that will sound your death knell.

I remember following shiny snail and slug trails across the concrete sidewalk when I was small. In this recent encounter, the slime trail I sought, but didn't find, and the earnestness of my particular specimen, reminded me of some stunning phrasing penned by Rick Bass. He was writing about tracking, too, albeit about animals in the snow somewhere out West, but different territory, species, season, and climate notwithstanding, his words resonated as I viewed my specimen on its slow-crawl mission: "I continue to look at all the other various tracks and realize that I am learning what so many others before me have learned: that there is no sense that can be made of it, and that it is more frigid and painful and hollow than you ever dreamed it could be, and that you want to lie down and quit but that because you are hers you do not, and you keep going."

I take these words into my core. For me, the "it" is the sorrowful aspect of my life, and the "her" is my mother, who is leaving me in gradual increments. The coming weeks entail a move to the dementia unit at her assisted living facility, where she can be watched more carefully and closely. There is a bleakness, an awfulness to this, even in the face of my gratitude that this care is available to her, that we were able to sell her house to finance this.

It is a big and scary world for a mystified child of any age who needs to decide for her mother, who needs to plan for a hospital bed and moving furniture and where the TV will plug in, when what looms above it all is a large and significant departure that will unavoidably hurt, no matter the preparation. But because I am hers I keep going.

For now, it is high summer. Whatever is looming, we'll eat by the river. We'll savor the long days and lush greenery. We'll take it all into us, every moment, every brush of contact with the fruitful, relentlessly regenerating earth. We'll make our way until we can no longer keep going.

How rich is the soil of this world, that a slug can lay these comforts directly at my feet?

Childhood Among the Ferns

The word "fern" sounds to me like a green promise, like a cool, familiar bed tucked away from midday heat. Silent, ubiquitous roadside sentinels, the ferns witness our early morning parades of beagle pulling woman, then the reverse on the way home here in Deep River. Some are a dark, forest green; others approach the neon end of the chlorophyll spectrum. Nature writer David George Haskell calls them "shade specialists"—ferns thrive quietly in the cool coves formed by larger looming plants.

In most bouquets, ferns play stalwart supporting roles for "leading ladies" like roses and daffodils. But this retreating style hasn't always been the fern's signature persona. In the Victorian era, a widespread passion for ferns engendered the need for a new word describing the craze: pteridomania. Lives were risked and felonies committed in pursuit of rare species. Enthusiasts kept fern albums detailing the many varieties they'd acquired. They indulged in diminutive hothouses, called Wardian cases after English botanist Nathaniel B. Ward. Some signers-on to the hotbed of enthusiasm were only using ferns to mask more lascivious intent: men and women in an otherwise restrictive era had the excuse to pair up in wooded seclusion during society-endorsed "fern-hunting parties."

A passionate alliterative phrase penned by photographer David Nicholls told me that pteridomaniacs live among us still: "There are few things more beautiful than a fern frond unfurling." The poets agree; the plants linger under the eaves of their verse. Dylan Thomas wrote about his childhood in "Fern Hill." Ted Hughes' unfurling fiddlehead sets in motion the workings of the entire world in "Fern." But it is Thomas Hardy's "Childhood Among the Ferns" that moves me, its protagonist literally enclosed by tall, benevolent plants that protect him from the rain and make him want to stay forever apart from the world of adult responsibility:

I sat one sprinkling day upon the lea,
Where tall-stemmed ferns spread out luxuriantly,
And nothing but those tall ferns sheltered me.

The rain gained strength,
and damped each lopping frond,
Ran down their stalks beside me and beyond,
And shaped slow-creeping rivulets as I conned,
With pride, my spray-proofed house...

This spring, our family listened to the *Peter Pan* audio book, sparked by Gavin's trepidation about the inevitability of growing up and what losses that might mean. It makes me wonder what impression my grownup self has made, because he seems to anticipate the loss of fun and freedom, the taking on of a large and heavy yoke that makes him want to stop the birthday clock at eleven.

He fashioned himself a fort from the sheets on my bed this morning, and later we did some fern gathering, trying to figure out if all of our specimens were true members of the species, or if some were "fern allies," lookalikes with spores concentrated in one place instead of in neat little hedgerows repeated on every leaf bottom.

It was a good morning; a memorable one. I am thankful for the wisdom that reminded me to reintroduce TV-free Sundays. We've stretched out time with our puttering; the world has opened up to us. I want to camp out on this deck and wile away the hours until again I must dress for work, putting away my meditations and musings on the green and growing in exchange for more orthodox endeavors.

Being a parent challenges me. My son, like all grow-ing sons, can be insolent, egocentric, and irritatingly silly at turns. But when I see him romping in the summer rain of our yard, examining the bare stumps that were until, just recently, trees; when I watch his brow furrow as he pores over our rapidly curling assemblage of picked spec-imens, I am with him. I am with this person who takes in and shapes the world even as he longs for it never to change. Like Hardy's tall fronds I want to lean over him and protect him—from growing pains, from the frequent stress that goes with being a grown-up human, from having to navigate in a world that can be so unwelcoming at times. Soon sixth grade will be upon Gavin, and with it some measure of social unease, and unpalatable and impractical academic rungs that must be climbed.

This morning when I walked alone, some musical part of me was aware that it was Sunday, although lately church seems a far-off call. I took to whistling hymns, and again found myself drawn to the layers of comfort I find in "This Is My Father's World." It reminds me of the wordless music that I hear in nature, and that it is only in "listening ears" that the full resonance of the gift can ring:

> *This is my Father's world,*
> *and to my listening ears*
> *all nature sings, and round me rings*
> *the music of the spheres.*
>
> *This is my Father's world:*
> *I rest me in the thought*
> *of rocks and trees, of skies and seas;*
> *his hand the wonders wrought.*
>
> —Maltbie D. Babcock (lyrics),
> in *This Is My Father's World* (hymn)

The tune blowing shakily from my lips is also quite a literal reminder of my father, gone since I was six but forever tied to all of my time outside. I remember squatting to feel slick blades of grass with him on the front lawn, and hearing him explain the concepts of dew and fog (he described the latter as "clouds on the ground," which enchanted me). I remember our car pulling off some Vermont mountain highway so we could gaze down at a beaver dam, and peering at countless cows through countless wire fences together. Did he know what a gift he was giving me then?

I know that as a parent I cannot be that tall grove, like the one in the Hardy poem, into which my son can escape, eschewing the demands of adulthood. No matter how protective my impulse, the world is waiting for him and my fronds are thin. But I have more than just a glimmer of hope when I see that he instinctively returns to the well of the world. He has developed his own gratitude for the gift, and with it protectiveness, for the "rocks and trees and skies and seas" that have immeasurably enriched his brief life. So far, his affections for the natural world have found their way into the farm club at school and countless Boy Scout trips and classes, and the desire to be a naturalist. Whatever the world brings forth, Gavin will have the language of things green and living to ponder and decipher—to open him, fortify him, feed him.

The ferns, I pray, will continue their silent watch, steadfast as he grows up alongside their unfolding generations.

Ants, Plants, and Pescetarians

When Gavin was in second grade, we indulged his fervent and longstanding request for an ant farm. We'd seen them in catalogues over the years, and seven seemed like a ripe age for acquiring one. He was old enough to understand that ants, as one of the ant farm seller's FAQ pages quips, "do not like earthquakes." In other words, they cannot be shaken; they must be carried with great care. They should also not be heated, floated, chilled, thrown, teased, painted, poked, or subjected to any other such humiliation or abuse.

I was not prepared for the horror that this Plexiglas prison would inflict on my psyche. The ants lived without a ruler, since federal law prevents shipping or selling the queens. They ate the gel in which they created tunnels and established traffic patterns. They ran about, seemingly with a sense of purpose but with no hope of propagating their kind, munching their way through the chemical goop that was their home.

As comrades began to die, the survivors sequestered the corpses in a special room they dug into the matrix. Water and "normal" ant food wasn't necessary (the gel—shades of SoylANT Green, anybody?—is designed to fill all nutritional needs). There were no intruders, but the price for that was no variety, no climate, no exit. Eventually everybody died in the stale air. I still get wildly claustrophobic thinking about that "educational" display on my little boy's bedroom dresser. I couldn't wait for the grinding descent to come to a merciful, motionless close.

I learned later that we could have opted for a sand-based ant farm, the old-school option. At least in this scenario water and crumbs of bread are occasionally introduced, making it more like solitary confinement than being encased in an edible casket that's entombed in glass through which other, more fortunate beings can gawk at you.

It's obvious that my ant farm recall can't be summoned without also calling up an escalating rant. I'll just say this: I hope those ants were capable of bringing each other comfort in that hellhole. Because it was a stark reality, regardless of the all-you-can-eat, all-expenses-paid glutinous buffet, regardless (or perhaps because) of the fact that the prisoners' life span was on par with that of ants in the wild.

One of the true pleasures of my own childhood was squatting down to observe ant colonies, little sand condo complexes that sprouted up between sidewalk cracks. What at first seems like chaotic behavior, basically a lot of herky-jerky ambulation, takes on some sense of order with closer observation. A lone forager ant alerts her coworkers that there's food, and before long there are many on site, gathering nutrition for those relatives who don't forage, the many mouths asking for food from within the catacombs below the hill. Other ants can be seen carrying waste away from the doorway. Dead ants are brought to the garbage heap (called a midden) along with the day-to-day trash. Scientists have recently proven that carpenter ant roles change in a predictable order as they age—from nurse, to cleaner, to forager. It's not clear what sparks the job shifts, exactly. The ants are keeping quiet about it.

Gavin and I have a ritual of putting a spoonful of sugar out for the "wild" ants, on our blue slate staircase landing where he also drew sidewalk chalk frescoes. The best aspect of our ritual is watching the mound of sugar for a while, wandering off when the actionless scene becomes intolerable, and then accidentally coming upon it again later, the white pile entirely obscured by ants who are partying with gregarious and grateful gusto. We are the source of manna from heaven in that particular slate-step universe, partial penance for the ant farm from the ninth circle of hell.

I started this chapter because Gavin heavily campaigned for ants as a topic. One night I listened to Harrison Ford narrate a *Nova* episode about E.O. Wilson called "Lord of the Ants," and I jotted all kinds of notes about biodiversity and ecosystems, ecological refugees, and Wilson's big, eager plan for the ultimate *Encyclopedia of Life,* a tome to catalogue every known species. I started plotting an impressive piece on the hidden lives of ants and what they do for our larger world.

But then Gavin inexplicably pushed an agenda on something else. We were surfing our new "Docurama" channel and Gavin kept asking if we could watch a documentary called *Vegucated*. Quite the left-field request, considering I can count the vegetables he likes on three fingers (chick peas, mushrooms, and scallions, but the last one exclusively in savory pancakes).

So we watched *Vegucated*. It was a kind of experiment, in which the director took a sample of open-minded, omnivore New Yorkers and encouraged them, using a fairly structured program, to become vegans. They dug around in their refrigerators; they visited a health food store; they learned that some ordinary packaged foods are in fact vegan; they had some spicy ethnic meals that didn't seem any the worse for lack of meat or dairy.

But, of course, vegan is a pretty hard-core commitment if that's not where you come from. A Latina girl whose dad cooked all kinds of delicious meat had to buck centuries of South American tradition. A single guy who'd been hanging on to a rack of lamb for a special occasion had to keep his plans (literally) on ice, and he was pretty sure his parents thought he had joined some freaky cult. A single mom was no doubt wondering how her kids might fare with this radical shift. There was much hemming and hawing among the well-meaning but conflicted experimental group, and I could relate. I've done my own share of hemming and hawing on this particular subject.

So, in quiet but firm fashion, the vegan director took her little group to a movie about animal factories. This short-form reflection doesn't have room for another full-blown rant, nor is it the place to house a litany of shocking statistics. But here's the thing: the soul-sick feeling I got about the clipped-beak farm chickens so fat that they couldn't stand and the cows continually inseminated and continually birthing, each calf taken within 48 hours, reminded me of that blasted ant farm. Eating meat from factory farms reminds me of how I complicity condoned the ant farm, just by having it in my home.

When I look at the issue squarely, I can't find a rationale for regularly abusing animals before they even get to slaughter, or for the copious spilling of water and grain to make bigger cattle when our earth is hurting and one in eight of us, globally, is chronically undernourished.

It is fitting to invoke the ninth circle of hell when referring to ant farms as well as factory animal agriculture. This is the circle of treachery—"fraudulent acts between individuals who share special bonds of love and trust." It's fraudulent to pretend that industrial animal farming is not rife with systematically cruel practices. Those of us privileged enough to make choices about our food are faced with some decisions.

The reality is that my hectic, working-mom life is not compatible with a lot of homework scrutinizing all meat sources, and veganism at this point sounds like another full-time job. So for now I'm pescetarian, which always sounds to me like another Protestant offshoot. I eat fish but no meat, and I am hoping to wean off fish, too. Maybe being vegan is on the distant horizon. My newly Buddhist husband is joining me, since he's having trouble resolving carnivorous tendencies with the ideal of "respecting and aiding all sentient beings." And Gavin, who's never loved a vegetable, is the most motivated of us all. He even manages an extreme feat: avoiding meat at the highly carnivorous Boy Scout campouts.

I am not a fan of fanaticism. I was first inspired to avoid meat when I read Jane Goodall's *Reason for Hope*. She describes, matter of factly, how she lost her taste for meat after learning some unsavory details about industrial farming. I admire Ms. Goodall, and given her intense, lifelong love of animals was at first surprised to read that she is not a full-throttle vegan—she just can't manage to find the right foods when she travels, so for practical reasons she sticks with the vegetarian route (allowing eggs and dairy). In fact, she goes out of her way to steer conversation away from a stringent set of dietary rules and towards pure and simple mercy instead: "…I want to make it quite clear that I do not condemn the eating of meat per se—only the practice of intensive farming. Let the meat eaters among us—most of my friends—try to partake of the flesh of animals who have enjoyed their lives and have been killed in the most painless way possible."

In *Twelve by Twelve*, William Powers writes that "everything comes from the earth. It's fine to grasp this intellectually, but to once again touch, breathe and eat this reality feels like reconciliation with a loved one after a long feud." Reconciliation, although an unfamiliar flavor, tastes good.

Close to the Edge

When I was a kid on Long Island I routinely cycled down to the end of Margaret Boulevard, past all the square lawns and grey, symmetrical curbs to the cul de sac. Alongside the edge of the street ran a shallow, narrow stream that grazed a string of backyards. It sported a softly bubbling margin of foam. I liked to stand on the edge and look down into the water, thinking about its largely ignored journey past many homes.

Once I saw a big, toothy rat walking through the stream with impressive aplomb. It reminded me of the oversized white possum that had slid just barely under a parked car when it saw me making my way to the bus stop. Where did these creatures go on their daily adventures? Where did they make their homes? Sometimes I walked up the block to our "woods"—one square, undeveloped lot full of scrub

oak and the illusion of the wild, when you were in the thick
of it. And during our walk home from Camp Avenue El-
ementary, my friend Adrienne and I sneaked down an alley
to the rear of the squat, concrete Merrick Life newspaper
building and checked out the mismatched patio furniture
sitting amidst a modest thicket of suburban greenery. We
relished our discovered gem and imagined the workers
coming out on their breaks to enjoy a few gulps of knoll-
filtered air.

Now that I am one of those adult workers I imagined on
the patio, I fervently hope that someone made it outside to
that little green space from time to time. My own workplace
has a largely unused and weather-beaten picnic table in view
of my boss' office, just a short walk across a verdant, and
quite rarely disrupted, lawn. But time away from the desk
is precious. I usually ignore the table and instead follow the
edges of several parking lots to the grassy strips behind the
buildings, looking for signs of life.

The first time I tried to circumnavigate the business park
perimeter, I made a wide detour when I happened upon
some zealous hornets favoring one building's foundation.
Then I watched from a little distance, wondering about
their intents and purposes as they plunged industriously in
and out of holes in the concrete. Another time I spotted a
deer hightailing it out of the lot, a cousin of the three I had
seen down the hill. After lunch one Friday, a large raptor
of grey-brown, etched magnificence impressed me with its
swooping flight onto a high branch overlooking the parking
lot. Its incongruity with blander surroundings had me
perceiving it as a brief and bewitching hallucination. The
day I created a birch-limb ramp so a stranded raccoon could

climb back out of the dumpster amounted to a banner lunch break adventure. I watched from afar as the scrawny captive ascended, peered cautiously at me, and hastily returned to freedom in the roadside woods.

It's the contrasts that have me treasuring these moments—human and other, tame and wild, confining office walls and expanse of fresh air, squared off and rambling spaces—the side-by-side existences pepper any given day with a welcome, and heightened, sense of awareness. Something inside me becomes alert, riveted when I witness these juxtapositions. We humans notice what is different, the thing out of place.

We humans are also prone to longing. Charles Siebert, in *Wickerby*, describes our race as, "the only ones who long to be a part again of that to which we already belong." Our analytical natures and our words are gifts, but they also set us apart from the rest of nature. We look past our boundaries for "the other" because it is not wholly other. We crave reconnection.

But aren't we, despite our distinguishing high rises, our tangles of wires, our myriad and complex neuroses, nature, too? Like the lions and gazelles on the African plain, our eyes, too, seek the horizon—that line between where gravity drops us and the untethered elements of the galaxy—when we look out across any expanse.

When my son Gavin was a very young infant, I confessed to the doctor that I worried about his mental status—he didn't look me in the eyes nearly as much as I'd anticipated. It was explained that my dark hairline against my pale, Irish skin was probably much more compelling to him visually. He was drawn to the "fence" on my forehead, where the sky of my hair met the lower reaches of my landscape. He sought out my edge and, for the moment, until his eyes and brain developed more fully, I was his horizon.

In *Last Child in the Woods*, Richard Louv talks about that inborn gravitation toward the edge: "… Research suggests that children, when left to their own devices, are drawn to the rough edges … the ravines and rocky inclines, the natural vegetation. A park may be neatly trimmed and landscaped, but the natural corners and edges where children once played can be lost in translation." This intuitive attention may actually have an important role in sustainability and survival. In nature, edges are places of transition that encourage diversity among the species.

Bill Mollison, in *Permaculture: A Designer's Manual*, explains that there is more variety found in the boundaries between ecosystems than in the middle. In a column for the Nature Conservancy magazine, a budding birder was surprised when her mentor took her to a power line clearing—the meeting of large zones of grass and scrub with the adjacent tall hardwood forest translates into a prime spot for spotting a wide array of birds.

The concept of permaculture is one of coming together across the edges—individuals with the community, and with land and space. There is more coming together overall in day-to-day American life lately, with an approach that includes shared cars, shared houses, shared work spaces, and a wealth of shared information and technology. This development, born in part of economic limitations, can have the fortunate side effect of being good for the natural world as well as human relationships—more sensible, modest, and collaborative choices can conserve resources while at the same time uniting us. Maybe there is more hope for a wiser brand of shared living now, because things are such that our edges are more likely to intersect and overlap on a day-to-day basis, like so many colliding Venn diagrams. Maybe we can become more connected and sympathetic overall, if only in small and imperfect increments.

In her autobiographical *Reason for Hope*, Jane Goodall writes about contemplating the work of doctor-turned-philosopher LeCompte DeNuoy. He suggested that we are evolving slowly to acquire moral attributes that are more caring and compassionate, less aggressive and warlike. It seems this would require more exchange across boundaries, fewer hard lines in the sand (or at least, more walks across them).

As with the land, each one of us has our hard edges, our own defined boundaries, and we need to visit them with respectful regularity, honor them as special places of reverence and reflection. When we come together in the middle, we are better for them.

Wider Ventures and Wise Teachers

During the summer of 2015, I got to live for a week at Trail Wood, the former home of the Pulitzer Prize-winning nature writer Edwin Way Teale, in Hampton, Connecticut. This 168-acre memorial sanctuary, which Edwin and his wife, Nellie, named Trail Wood, arose from former farmland. It encompasses forest, meadows, ponds, streams, and a network of winding trails.

Edwin's office in the white Cape Cod house is preserved just as he and Nellie left it, filled with an impressive collection of books and mementos from their lives that they largely spent outdoors. Edwin died in 1980, Nellie 13 years later. Their haven—one that they longed for and sought for years—has been a Connecticut Audubon property since 1981 and has hosted visual artists and writers every summer since 2012. Several pieces here arose from my time at Trail Wood, which I view as nothing less than sacred.

Long gone from this earthly plane, Edwin is still teaching us about the natural world, its broader truths, and how we can help it, and so are many other wise and generous folks—fellow lovers and studiers of nature who have parlayed their own, rich experiences into opportunities for loving and caring for the world. This section reflects on time at Trail Wood, as well as other experiences and thoughts enriched by those who honor the gifts given by nature.

A Curious Path

July 27th arrived with an auspicious dawn chorus. Although only about an hour from home, I heard new bird calls and wondered from which species they emanated. I jotted notes and sipped coffee. It was my first morning awakening at the Trail Wood Memorial Sanctuary, the former home of nature writer Edwin Way Teale and his wife Nellie, in Hampton, Connecticut, where I was the writer in residence for a week.

I started my wanderings at the insect garden, remembering Teale's history of intrepid insect observations and photography. The thicket buzzed busily. A small wasp landed on my hand, toting what I think was its meal (some kind of inchworm or larva). It seemed oblivious to me and flew off when I finally shifted my arm. A bit later I walked by a rustic wooden bench under an arbor, and it recalled the stick blind where Teale liked to cover up and wait for birds, mice, and other small creatures to climb onto or into the twigs, parlaying his invisibility into a perfect close-up observation opportunity.

I stepped down to a flat stone in shallow Hampton Brook, bending low and peering at a cluster of dark minnows, and then at whirligig beetles swirling in the water. From my perspective, they looked exhausted, swimming in ceaseless and frenetic arcs. I recalled a treasured book, *The View from the Oak,* by Judith and Herbert Kohl, which introduced me to a shorthand term for "the world around a living thing as that creature experiences it." The word is *umwelt,* and I wondered about the whirligigs' umwelt. Of course, they have a demanding job. These driven creatures keep the insect population checked, also clearing the water of dead and dying insects, all the while gyrating frantically (surely giving fish and bird predators a run for their money).

Later I watched a hummingbird clearwing moth. It took me a few long stares to make sure that's what I was seeing. Like actual hummingbirds, they move quite rapidly, and in an attempt to be sure I researched whether the moth or the bird likes wild bergamot, the flower around which my find was hovering. I'm not sure who's mimicking who, because the "original" bird and its insect "copy" both go for bergamot nectar.

So many of my notes begged questions. What is the life of this creature really like? What motivates it? Why does it behave as it does? In the case of the hummingbird moth, is it ever fooled by an actual hummingbird? And then I had a question about the questions: What does all of this close-up observation, all of this questioning, teach us humans, beyond the simple acquisition of scientific fact and perhaps learning how it can work to our advantage?

When I read one of Teale's late July entries in *A Walk Through the Year*, published nearly 40 years ago, I felt affirmed when it transmitted his many questions. He was already so knowledgeable and observant at that point, having published about two dozen books before this one, but he noted: "Again I am encountering something new beside the pond." He went on: "The whole bottom of the shallows along the dam, I notice, is sprinkled—as with a dense scattering of seeds—with tiny, flat, rounded objects." He then described his investigation with Nellie, complete with 14-power magnifying glass: he had found pill clams. Then the questions: "Why this sudden population explosion...? Why on this particular summer?"

Curiosity isn't unique to humans. I am always amused by the resident carpenter bees in our garage, who fly up to me when I exit my car, clearly sizing me up and zooming in even closer for further inspection. But as far as I know, carpenter bees don't research us, keep files on us, or get together in clubs to talk about when they saw us and how we are doing. I don't think they're trying to protect us from climate change either. The human variety of curiosity is much more likely to lead to compassion in action, one that can offer some tangible help.

When I first arrived at Trail Wood, happenstance allowed me to join a small writing workshop led there by Alison Davis, a spry elder who described herself as a bridge between the Teales—both of whom are now long gone—and modern day. The Davises and their older nature-loving neighbors grew to be devoted friends who shared picnic breakfasts together.

Alison had in her hands a speech that Teale had written. Within it he talked about nature being the most immediate extension of the beautiful and the good. But he went on to muse about one thing he observed as mostly missing from nature—compassion! Of course, there are exceptions, including the patient and kindly way in which many animals raise their young. And Teale didn't have the immediacy of mediums like Facebook, where I just saw a leopard taking care of a day-old baby monkey, the footage all the more interesting because it had killed the creature's mother only minutes before. But I got Teale's point immediately: the weak, the lame, and the smallest are often left behind in the natural world, to fend for themselves and often to be preyed upon.

Humans have a unique calling to be compassionate. Albert Schweitzer said: "Only in thinking man has the will-to-live become conscious of other will-to-live." The obvious, age-old contradiction of human cruelty doesn't negate myriad examples of humans reaching out to others (human or otherwise), offering help, sometimes even losing life or limb so that another—often weaker—being can continue to thrive.

Experts in the documentary *I AM,* a film looking closely at human purpose, talk about the biological response we have when feeling deep compassion for another—tears well up, we get an expanded feeling in our chest and a kind of rush. We often actually wince when pain is inflicted on another. We are, if you will, living another's umwelt for that moment.

I AM expounds on a conclusion by social psychologist Jonathan Haidt: we humans are "hard wired for a compassionate response to the trouble of others." Later in the film, evolutionary biologist Elisabet Sahtouris talks about an idea that continues to gain scientific ground. Research keeps bringing forth ideas of interconnectivity and our key role in connecting with all other species. Of course, this isn't a new concept in the contemplative realm. Schweitzer wrote: "A man who possesses a veneration of life will not simply say his prayers. He will throw himself into the battle to preserve life, if for no other reason than that he is himself an extension of life around him."

Alison's class came back to the Trail Wood picnic table with notes on what they had noticed—the flight of a bird, the path of a butterfly, the clouds, a fallen tree. The inevitable questions came, too: What kind of flower was that? What was the woodchuck eating? Were there really fewer butterflies than before?

There was childlike joy in many of these musings, but also a studious and purposeful air, a sense of deep concern that reminded me of an interesting detail I had stored away: Biblical scholar Ellen Davis recounted a story about her teaching assistant helping her prepare a test on the Old Testament. He suggested that she include questions about land, because it came up all the time in class. This surprised her. Despite her scholarly work, Professor Davis hadn't yet tuned into the fact that the land is mentioned constantly in the Hebrew Bible: "land, water, its health, its lack of health, the absence of fertile soil and water."

She realized that she couldn't read the Old Testament for very long before another reference to land cropped up, and came to recognize "a huge gap between the kind of exquisite attention that the biblical writers are giving to the fragile land...and the kind of obliviousness that characterizes our culture..." We humans have lost our way, lost sight of the land in so many ways. Of course, the land is the matrix on which so many of the creatures that incite our curiosity and compassion must thrive. Any vigilance and action, any protective impulse we direct toward the land and its inhabitants is an outgrowth of the dual, and very human, gifts of curiosity and compassion.

Often, in the midst of my dizzying life as a working mother, I feel guilty that paperwork and housework are pushed aside so that I can squeeze in a long walk in nature. I wonder about the impression I make at work, too—I often apply my makeup and make my hair presentable after I arrive at the office, having chosen to use my morning time to be outside rather than to groom to a professional standard. In more than one staff meeting, I've had to surreptitiously remove a bug who'd hitched a ride from the forest with me. But thinking about it more, I am sure in the knowledge that these long walks are anything but self-indulgent. They are necessary. In fact, to me they are mandated, as is the alertness and curiosity and compassion that they help to engender.

One early morning, I took a favorite route through my home town of Deep River. I parked at the IGA supermarket lot and walked past a string of historic homes, an old cemetery, and a small cluster of condominium dwellings before I stepped over the railroad tracks and arrived at Pratt Cove, a freshwater tidal marsh. As I peered past the thick grasses, looking for water birds, a curious ripple caught my eye. Something big moved just below the water's surface.

To my delight, a river otter emerged up onto a bank, its curved, wet body lithe, its mouth holding a large fish. I fiddled with my camera buttons, already knowing I wouldn't be fast enough to capture it. But this image remains fixed for me, no photo needed: the minute it rose out of the water, for maybe three seconds, it looked directly at me. I looked back, wondering what it was thinking. Then it took its breakfast into an unseen place within the reeds.

My curiosity whetted, I researched the otter, rewarded by descriptions of its childlike playfulness—tossing pebbles around and catching them, using stream beds as water slides. This behavior isn't limited to the young. The otters are completely engaged with their surroundings, seeming actually enthralled with them, lifelong.

There's a reminder for us here, to retain the best qualities of childhood: the eyes-wide-open curiosity and compassion that mark the best aspects of humanity from a very early age.

It starts with a walk, or a long period of sitting and watching the denizens of the natural world going about their daily tasks. From there, from this curious place, we will begin to know what we need to do.

Beckoning the Mountain

I have always believed in conquering the mountain, nodding in agreement with Francis Bacon, who famously proclaimed, "If the mountain won't come to Muhammad then Muhammad must go to the mountain." Proud of my many thoughtful treks over and around New England hillsides, I have gone to the mountain a lot and plan to continue the habit.

I know that my journeys have nourished a sense of connection with the Earth. But I have also started to heed an insistent perception that I must sit still more often, applying a counterweight to the nearly ceaseless motion that underlies my working and family life.

As a writer in residence at Trail Wood, I had plenty of trails to cover this past summer. But I also, at long last, had plenty of time to sit and look and listen and think. I didn't have to go to the mountain. It was all around me.

In *Flat Rock Journal*, Ken Carey writes about "Moments when my awareness recognizes itself in all I see, and every pebble and leaf and tree looks back at me, mirroring some facet of myself." I know from experience that this kind of awareness can be stamped out by forward thinking and determined feet. But being still—both physically and mentally –has largely eluded me. Even slowing down is a challenge.

At the Trail Wood nature writing workshop with Alison Davis, we students were sent off to wander the grounds, and I so admired the thoughts that Alison shared when we rendezvoused to discuss our impressions. She had not ventured far during that hour. The spot she chose for her nature time was the picnic table on the lawn behind the house. She described the initial, antsy observation that not much was happening; how perhaps she should move to the woods, where there could be something more exciting. But she stayed and she watched and listened, and the variegations in the grass' color started to reveal themselves, as did bugs and birds and a lively breeze carrying plant scent and a hint of weather shifting.

Alison shared an anecdote about Nellie Teale. She recalled that it could take Nellie an hour to walk around the small pond near the writing cabin. Nellie would dip a white ladle into the water again and again, fascinated by what she found with each tiny harvest. I loved the thought of Nellie's cup overflowing with the fruits of her attentive stroll.

One morning at Trail Wood, I decided to experiment with being more Alison-and-Nellie-like with my outdoor time instead of taking off down one of the many paths through the woods. My plan was to simply watch and listen and perhaps sniff, all from the vantage point of the picnic table. I had to set a timer so I wouldn't keep checking my watch. I planned a half hour facing the house, where I assumed the woodchuck would entertain me for at least part of the time, then a half hour facing the opposite direction, a space chock full of brush and older trees and tender plants encircled in deer fencing.

It wasn't easy to go cold turkey from the need to foot it. I allowed myself to jot some notes at first, giving my hand a task to keep it occupied. Shadows of birds passed over the lawn, and I heard their owners' calls a beat later as they found perches deep in the trees and shrubs. The resident woodchuck waddled around near the stone wall, keeping one eye on me at all times. A distracted yellow butterfly thwacked into my forehead, and I chuckled as he hurried on. Birds hopped across the grass, and it struck me that the woodchuck and the birds were doing their daily work – albeit on unhurried missions—in serious pursuit of their respective plant and bug suppers.

When the timer went off to signal my second stillness session and I turned away from the house, it dawned on me that the habit of scanning the landscape for unusual figures or objects likely evolved to support human survival. But I noticed that looking and listening with relaxed and patient eyes and ears was far more fruitful—in terms of tapping into a sense of connection—than my usual habit of surveying for a sudden burst of color or movement. Instead of waiting for something "unusual" to happen, I started to closely attend to whatever happened to be around me. It didn't take long to notice a gold beetle climbing a tall blade of grass, and then its brother two feet from it, right next to my shoe. The second insect used my foot as a bridge to the next green spire.

After a while I trained my binoculars on the middle distance. A hoary tree tangled in a berry-laden Virginia creeper was clearly the hot spot for the local avian community. The stars in my sights were a sweet pair of bluebirds. The male darted to and fro, stopping to feed a quick treat to the female. After some restless-looking hops he moved to be closer to her and, as if posing for a John James Audubon tableau, they turned to face me with regal postures, the male's head just a bit below the female's.

I took my growing skills in thoughtful, unhurried observation for a walk around the pond, trying to duplicate Nellie's practice of unhurried circumnavigation, dipping a white cup into the water. In one cup, a thriving water insect. In another, a remarkably detailed husk of one—perhaps the molt of a growing bug? The paper cup turned greener with each scoop, accumulating a coat of algae paint. The rhythmic sound of pouring water seemed a kind of mantra for slo-mo contemplation.

I doubled back for a stroll down Veery Lane, working on moving slowly and with deliberate use of my senses. My novice walking meditation was rewarded by sighting two turkey hens and their poult charge on a low perch by the pond-side screen house. They clucked in annoyance at my intrusion, so I sat partially hidden behind the screen and ate an apple, musing about the little family and its water view as I listened, looked, and finished my snack.

When I got back to the Teale home I was drawn to a book of Mr Teale's that I hadn't read before: *Days without Time*. Teale's words rang just as true more than 60 years later: "The fall of the tree, the swoop of the hawk, the tilt of the buzzard in a windy sky, the song of the hermit thrush at evening, the opening of a windflower, the eddy of a wood-land brook—all of these are events for days without time... Ticking clocks and factory whistles have little to do with the eternal recurrence of these eternal themes." This was a message I needed to hear. So much of life can be driven by clocks, if we let it. With practice, being quiet and still accesses a deeper dimension, one that is always there but is often untapped, one that eclipses the metronome of recurring obligations and to-do lists.

I visited Trail Wood's Beaver Pond daily, and I noticed that the beaver finally made an appearance when I had stopped looking at my watch, when I stopped wondering intently if he would appear within my hour seated on the bench. From there my mind jumped, tangentially, to many stories I've heard about devoted couples meeting when one or the other had stopped looking for the perfect mate.

It's easy to wonder if there is some magical cause and effect involved in this equation of "not caring" in order to get the desired outcome. But the simpler conclusion is that there is no equation involved at all: when we can put our agendas, timepieces, and deadlines aside, we find that the world continues to unfold anyway, often in ways that surprise and delight us. Sometimes, it's good to let the mountain do the work.

The Way in the Woods

A Week with Edwin

The summer before last, a bookstore owner in Woodstock, New York, listened to me talking excitedly about Trail Wood and the residency I was hoping to win. This spurred him to climb up into his attic and rummage around, emerging with a slightly musty but well-preserved four-volume set of Teale's *The American Seasons*: one delicious book for each season, chock full of Edwin and Nellie's travels around the United States and expertly crafted prose about the natural world. *Wandering Through Winter,* the final book in the set, won the Pulitzer Prize for General Nonfiction in 1966.

The set includes a thin biography of Edwin, and in it I was tickled to read that he declared himself a naturalist at age 9 and by age 12 had changed his middle name from Alfred to Way. Edwin felt his full given name was "too commonplace for a future Thoreau" and instead adopted his

Grandfather Way's last name as his middle. This boy already knew he would immerse himself in nature and write about it. And, an inspiration for any writer, he found a way to do it full time. Every year he remembered his personal "Freedom Day," the day he broke away from a salaried magazine job and started his freelance photography and writing life. When he began his time at Trail Wood he was, as they say, living the dream.

Sadly, most of Edwin's 31 books, reflecting a unique and riveting mix of accurate science and genuine, childlike delight, are out of print. Many friends, upon hearing that I'd won my residency, followed their hearty congratulations with a short pause and then a polite question. Who was Edwin Way Teale, exactly?

When I pulled up the long Trail Wood drive to meet Rich Telford and Vern Pursley, the residency coordinator and property caretaker who gave me my orientation tour, I soon made my own short pause followed by a polite question. They advised me to drive around the bend and park under the catalpa tree. I paused; my mind whirred nervously: I had no idea what a catalpa tree was. By asking them, was I about to reveal that I was not a "real" naturalist?

I've read enough about and by Edwin Way Teale to feel comfortable calling him by his first name, and also to assume that he would not have judged or berated me for my limited knowledge of flora and fauna. His writings reveal a deep sense of unabashed joy and enthusiasm for observation and learning. Many people wrote to him asking about the natural world and his books, or asking to visit. His responses were kindly. I've detected no trace of arrogance despite his obvious expertise. His zeal lives on at Trail

Wood, embedded in the ripples in the waterways, the insects' buzz, the woodchuck perhaps descended from the very one he wrote about, and the big, bean-like pods of the iconic catalpa that graces his yard.

My time on the preserve was rife with questions. The sight of even a common gypsy moth had me peering and tracking and researching in a sort of timeless bliss, one question leading to another and another. I studied Edwin's correspondence (meticulously preserved and housed at the University of Connecticut's Thomas J. Dodd Research Center) and found the same impulse to question and learn in letters to his friend and fellow writer Rachel Carson. He wanted to learn more about eelgrass and sea anemones, asking Rachel for her help in this quest. In another letter, Rachel recalls their discussion about why the conch shell seems to give off the sound of the sea.

Perhaps as a reaction to my experience at Trail Wood, which was as contemplative as it was active, I became very curious about Edwin's spiritual life. I found nothing on record to suggest that he seriously embraced a specific religion (at least not publicly). Recently, however, I heard an interview with social psychologist Ellen Langer that helped me give a word to the spiritual "vibe" I detect in Edwin's very deliberately chosen, very lively way of life, studious but also peppered with an enviable dose of delight. Langer's definition of mindfulness has little to do with meditation or yoga or prayer. It is "actively noticing new things," which "puts you in the present" and is "literally, not just figuratively, enlivening. "

One can notice new things anywhere, but I did so alone at Trail Wood, free from the usual constraints of work and family responsibilities. My week was a festival of noticing, intertwined with bouts of trying to capture it in words. My mind filled with things that enriched and enlivened me, reminiscent of Edwin's reflections on his own experience. *Mind-full.* I wrote about the hummingbirds and their favored roses of Sharon, the sphinx moth and its affinity for the catalpa, the multihued abundance of mushrooms in the woods, the catbird who flew around inside the information shed at dusk, and the beaver in the far pond who finally revealed him (or her) self early one morning.

The "Way" in Edwin's prophetic adopted name leaps to mind when I recall my week alone on his cherished preserve. I took a laminated map that showed me the way through Ground Pine Crossing and Fern Book trails, to the Beaver Pond and to the Hired Man's Monument. Edwin's words—which I read at the start of my day while sipping coffee on the low slate stoop of his house—showed me the way to things that are too easily abandoned by the workaday world. I found the way to silence and observation with all senses. I found the way to noticing and then noticing some more, and recording something of what I thought I could be learning. I joyfully snapped countless photos and spent languorous hours in a hammock. I trekked, studied, wrote, and edited. I learned a way of stepping back and finding meaning in all of it—even the bug bites and humid air and my achy limbs as I hiked yet another hour.

Edwin's life at Trail Wood, and my own weeklong sojourn there, may read as paradisiacal, maybe even escapist. But I think the true test of Edwin's words, my own scribblings, and a hallowed place like Trail Wood is how any of these experiences resonate out in the "real" world. This, too, is a mark of mindfulness, according to how Langer describes it: this finely attuned noticing, this present-tense, spirited way of approaching life is also sensitive to context. Case in point: this reality-bound paragraph at the end of *A Walk Through the Year*, a footnote to 365 delectable word-snapshots of creatures and plants and climate and seasonal cycles at Trail Wood:

> "As I come to these final sentences, I sit here wondering if a time will ever come when such a book as this will seem like a letter from another world. Will the richness of the natural world be overrun, and more and more replaced with a plastic artificial, substitute? The threat is real. And the outcome seems to depend on the wisdom and courage and endurance of those who are on the side of life—the original, natural life, the life of the fragile, yet strong, out-of-doors."

Parade magazine hailed Edwin as the "press agent for Nature." Surely we could use him now, someone to show us the way to more quiet and thoughtful observation, to more well-informed and truly effective conservation. We have his words, still so very relevant, as a guide. We have his place, Trail Wood, as a touchstone for reveling in the outdoors, for the spark we feel when we make contact with plants and animals and the elements.

After living at Trail Wood, I've decided that the best remuneration we can offer for places like it, and for people like Edwin and Nellie, is to join the ranks of "press agents," honoring these places with our time, transmitting our enthusiasm and joy, and acting on the protective impulse that their meaningful and mindful gifts engender.

Fountainized

Scottish-born John Muir is best known for his advocacy of American national parks. But those who know more about him are likely to know about his famous "ride" atop a tree during a wind storm. His voice is gleeful and fearless when he describes how the adventure started: "...when the storm began to sound, I lost no time in pushing out into the woods to enjoy it." He describes the storm as "holding high festival," and he joins the celebration wholeheartedly, climbing a Douglas fir and clinging happily for hours as the tree sways mightily in the fierce wind.

Muir seems 110% immersed in his experience. His words about his tree ride capture the wild play of light across the landscape, the mosaic of color stretching for miles, and a bath of myriad and distinct scents from the Pacific and from plants both lofty and low. Saturations like these seem to be where Muir found his God, described as one who: "flows in grand undivided currents, shoreless and boundless over creeds and forms and all kinds of civilizations and peoples and beasts, saturating all and fountainizing all."

I've shared that same sense of being "fountainized" in the elements that marks the legendary Muir adventures, although the scope of my own adventures pales in comparison. Once, when I was still quite a tiny child, an immense wind lifted my umbrella and my incredulous self an inch or two for the briefest of moments, gloriously free in unwitnessed flight. I still remember the intense emotions I had at 10 when experiencing an especially majestic stormy day in Vermont. Although I gradually became aware that embracing the wind and rain with abandon might garner judgment in some circles, even as an otherwise "tame" adult I've continued to experience moments that Muir might call "fountainized."

I cherish walks in drenching weather, all the better with a bare head. During my last wet trek, my husband came, unbidden, to fetch me, assuming I was miserable in the torrent instead of reveling in how it overcame me. He had no way of knowing that my literal immersion brought a sense of a "mystic sweet communion," to quote an old hymn.

I've never felt quite as enchanted with the cold. Surveying the leafless landscape robed in snow used to leave me with the impression that there was not much "action" to observe during the coldest months, beyond the parade of birds and squirrels back and forth to the feeder. But a winter read, *The Backyard Almanac: 365 Days of Northern Natural History*, tells me that bark beetle larvae are tunneling just below the surfaces of many trees, and tiny dwarf spiders are more visible in their stark contrast to the white mounds obscuring the grass. The birds are also much easier to spot on their perches now that the branches are bare. The book, together with Muir as a role model, has me bundling up and venturing outside, even on chillier days.

Twice I've spotted a muskrat in the frozen marsh; he swims under the ice to gather food. On my last walk by his spot I waded through thigh-deep roadside snow, hoping to discern if the new tracks on the pond were his. I could see that there was no impression of a dragging tail, and my follow-up research on tracks left me wondering if there's also a resident mink nearby.

During my last trek in the cold, I realized that spring couldn't be too far off. I rejoiced in my first sighting of "tree circles,"—those rings of snow melt hastened by the warmed dark wood as the sun finally starts to make a dent. As the sun hit me, I had that same expansive feeling that Muir often recounts, even though my adventure was a simple walk up the block, minutes from home. Swells from the Hymn of Joy sprang forth in my mind, accompanied by a chorus of early birds who were rehearsing for the upcoming spring season:

Field and forest, vale and mountain,
Flow'ry meadow, flashing sea,
chanting bird and flowing fountain,
call us to rejoice in Thee.

Thinking about the thrilling contact I have with sunlight and wind and snow and rain, I recalled the many baptisms I have witnessed—a symbolic kind of "fountainizing," with tiny dabs of water sprinkled and then quickly mopped off the baby's crown. I thought about my mother's recollection of her much more intense baptism at 14—the grand gesture of being pushed fully below the water, coming up completely soaked. But even following the briefest dampness on the baby's forehead, I've treasured the pure beauty of her attentiveness to the moisture, and to the momentary overshadowing by the bestower. Sometimes even a modest spray of water can be felt as a wonderful, wondrous gush.

We Could Be Heroes

If we were to churn out action figures of the naturalists, there would be plenty of larger-than-life heroes to cast: John Muir, advocate of national parks. Rachel Carson, fierce and pioneering environmentalist. Henry David Thoreau, deep thinker in the woods. Annie Dillard, modern woman in the Thoreau vein. Edwin Way Teale, rare winner of a Pulitzer Prize for nature writing. All of these action figures could come with colorful pamphlets about their accomplishments, also detailing the parks, prizes, and other forms of remembrance honoring their names.

Teale is my strongest naturalist influence these days. The legacy of his words was made even more real by my nature writing residency at his beloved Trail Wood this past summer. He was a precise and "sciency" guy—he jotted down highly detailed observations and conducted small experiments to learn more about what his beloved creatures would do. He had a great knack for marrying scientific fascination with genuine joy at what he was learning and also the joy of just being out there, surrounded by wildlife, however ordinary or diminutive the species.

Many have made careers from being even more precise and sciency, collecting information, making charts, and writing papers and authoritative guides. Some of these men and women have been laid as a sacrifice on the altar of their vocation. Richard Conniff, author of *The Species Seekers* and *Swimming with Piranhas at Feeding Time*, keeps an updated Wall of the Dead: A Memorial to Fallen Naturalists, on his blog, cataloguing on-duty demises of naturalists, such as:

"Akeley, Carl (1864–1926), naturalist-taxidermist for the American Museum of Natural History, age 62, while collecting mammals in the eastern Congo, of dysentery."

"York, Eric (1970-2007), biologist killed, age 37, by pneumonic plague after autopsying a mountain lion in the Grand Canyon."

"Tegner, Mia (1947-2001), ecologist of the kelp forests for the Scripps Institution of Oceanography, died, age 53, while scuba diving off California."

"Saunders, Hamish (1976-2003), an oceanographer, drowned, age 26, after being swept off the remote island of Branca Rock, in southeastern Australia, by waves that reached him at 45 meters above sea level. He was a member of a team of four studying the Pedra Branca skink."

Tragic tales, of course. But it's an educated guess to say that at least some of the naturalists on Conniff's Wall died doing what they loved.

I've been thinking a lot lately about the fellow walkers I sometimes see when I am out on my own jaunts. I see that they, too, have their own regular adventures out of doors, even if "being in nature" isn't their primary goal, even if they aren't living and breathing their love for the earth 24/7 like some of those remembered on Conniff's Wall. Maybe they are walking for fitness, or treating the dog to a happy excursion, or getting away from a cantankerous spouse. But surely, they, too, are coming upon eye- and ear-catching moments—moments as simple as the sudden, startling flight of a bird out from a curbside bush, or a pair of feisty squirrels running up an oak as if it's a spiral staircase, chattering the whole noisy way.

At some point, at least for some of my fellow walkers, moments of surprise and curiosity and pleasure on these walks coalesce to encourage some tangible measure of devotion, such that the casual stroller morphs into a full-blown nature lover, goes out of his or her way to be out in the elements, weaves chances for such encounters deliberately into the fabric of daily life. Poet Mary Oliver writes that "Attention is the beginning of devotion," and I've become increasingly appreciative of those who commit themselves to attend to nature, to be there, to see what unfolds.

When I looked up "attend" in the *Online Etymology Dictionary*, I was reminded that it means so much more than "show up." Showing up is just the first step. My favorite part of the lengthy etymologic summary, which starts with "circa 1300, 'to direct one's mind or energies,'" is this part:

"literally 'to stretch toward'...The notion is of 'stretching one's mind toward something.' " In my case, the stretching isn't purely intellectual. My soul stretches along with my mind when I am attending to the birds' dawn chorus, the scent of freshly fallen pine needles, even a lone, garden-variety ant hurrying along in its determined quest.

I listened to a radio interview with the soprano half of the Indigo Girls' singer/songwriter duo, Emily Saliers. She was talking about how she didn't understand the power of quiet and spiritual practice at first: "...what's the big idea about being a monk and going and being quiet? What does that do for the world?" I think that those who haven't yet started to attend in the *stretching* sense of the word, who don't yet feel especially devoted and are caught up in the churning, pressured pace of civilization, can feel the same about going out of their way to spend time in nature—What does stretching toward nature really do for the world? What won't get done when we are out walking, or lying on our bellies, swinging our legs lazily in the tall grass and cocking our heads to hear a songster hidden in the bush?

For every heroic naturalist action figure there are countless unsung, with no unique claim to fame. They might not inspire a figurine cast in their likeness but are devoted in their own ways, and it matters in the same way that those anonymous monks praying for the world matter.

These rank-and-file devotees are the people who, when I post a fuzzy photo of a bird couple to my Facebook page, rush to confirm that, yes, these are dark-eyed juncos. They direct me to helpful Web sites and respond to my questions on beak color. They share a tip about the best seed mix. They are as taken with my discovery as I am. They live for quiet but colorful moments such as these, stretching their minds (and souls) towards life in many forms. Biophiliacs, all.

Before Edwin Way Teale was the *famous* Edwin Way Teale, he lived in Baldwin, New York. This suburb on Long Island is quite close to where I grew up, a town on either side of Sunrise Highway with an extended necklace of railroad stations packed every weekday morning with Manhattan-bound commuters. Teale lived in an ordinary house and had an ordinary staff writer post, but he also managed to pay a lease for "insect rights" to the owner of a nearby overgrown farm plot, for the privilege of planting things and wandering around, observing and photographing insects. His publisher and biographer, Edward H. Dodd, Jr, describes how his activity there drew attention, concern, and bewilderment, at "a full grown man peering into a grass clump, or stretched out prone to watch an ant at work milking aphides."

Reading Teale's work, you don't get the sense that there was any real risk involved, other than perhaps the chance of being stung by bees, bitten by mosquitoes, or sprayed by an irate skunk. There are throngs of devotees who have never risked life or limb in pursuit of nature-related knowledge or exploration. They fall more into Mary Oliver's camp, with its motto of "My work is loving the world." They love this world alongside their jobs, their kids, their caretaking, and their unforgiving schedules.

They are everywhere. They are writer Dylan Tomine, who in *Closer to the Ground* writes simply but beautifully about time outdoors with his family, foraging and exploring around Bainbridge Island. They are Connecticut artist Paul Enea, whose landscape of hills and trees in variegated greens sits in a silver frame in my hallway. They are my son, Gavin, whose love for nature comes forth in every art lesson, thick strokes of paint capturing the beautiful untamed to which he constantly gravitates. They are the elderly woman whose name I do not know, who walks many miles, daily, all around my town and the next. They are me, my iPhone overflowing with technically imperfect snapshots of what I find on my walks and dictations about the flora and fauna I need to learn about. And they are the many people who may never write a book or sell a picture, or even be mentioned by name in discussions of nature and its appreciation. They are the faithful followers, embodying their love for creatures, for weather, and for wildness in quiet ways.

So we have a whole gamut of naturalists sharing the earth, from near superheroes and genuine martyrs to many meeker examples who are primarily focused on simple appreciation. But I've come to question my own meekness lately. As much as I revel in my humble but joyful day-to-day walks, as much as I prefer my quiet, anonymous writing office to a lectern or an ecology-minded protest, I also feel that the depth of my attending, my stretching towards the world, and the joy I find in this pursuit obligate me to step forward and do more.

I feel compelled by things like the report by the Intergovernmental Panel on Climate Change, which is made up of hundreds of scientists operating under the auspices of the United Nations. Joel Achenbach, a *National Geographic* writer, summarized a key theme of the report: "This one repeated louder and clearer than ever the consensus of the world's scientists: The planet's surface temperature has risen by about 1.5 degrees Fahrenheit in the past 130 years, and human actions, including the burning of fossil fuels, are extremely likely to have been the dominant cause of the warming since the mid-20th century. " I believe the words of the Panel's scientists, although a part of me would like to stick my head quite deeply in the sand when I read:

> "Conversion of natural ecosystems, a driver of anthropogenic climate change, is the main cause of biodiversity and ecosystem loss (high confidence)... Climate change will exacerbate future health risks given regional population growth rates and vulnerabilities due to pollution, food insecurity in poor regions, and existing health, water, sanitation, and waste collection systems (medium confidence)."

When I do the mental math and carry predictions out to the Nth degree, I come to the conclusion that the world is, quite literally, shrinking, in terms of livable habitat. The ice caps are melting; flood zones will drive many from their homes. Drinking water will become quite scarce in other areas. But despite this sobering information, I think that there is something quite promising about this period in human history. Because the shrinking that takes place in the *virtual* world can do us some good.

In her well-loved and lauded book, *A Wrinkle in Time*, Madeleine L'Engle wrote about children who travel through the fourth dimension to save their father, witnessing the Earth threatened by dark forces during their journey. The Internet, of course, is firmly planted in our three-dimensional world. But it does wrinkle time, in a way. Its circuits allow us to learn about an awful earthquake in Nepal and send a donation for aid within minutes. We watch videos that make us instantly sympathetic to a foreign culture or an animal species or a differing point of view. In the virtual model of our earth, trustworthy sources can inform us each day about endangered creatures, and companies abusing the environment, and clever solutions posed by scientists and concerned citizens alike.

Yes, we often suffer from information overload, and it takes persistence and savvy to sort through to the real story. But, technology, the very force that in many ways has damaged, and continues to threaten, our environment, is also rich with opportunities for solutions. For most of us the most immediately accessible and useful technology is encased in our devices—laptops, tablets, smartphones. And even those of us who fall largely into the appreciator—versus the activist or hard-core scientist—category can step up to actually do something to protect the world with these devices as our primary vehicle. Even the meekest can participate relatively anonymously online in monitoring and hopefully protecting the many forms of life that we care about.

This ever-widening doorway for global contribution is called citizen science, and our reports on the world at a click's distance makes data more inclusive, more accessible, and potentially much more valuable, because information is transmitted so much closer to real time—lending greater momentum for real action—than ever before possible. We can be part of this *meaningful* social media. We can record something that matters, giving scientists a much greater reach than if they tried to do all of the research on their own.

My writer friend Melissa Gaskill stumbled into citizen science while in Baja California Sur with her three kids.

They volunteered to patrol a stretch of beach on the East Cape for an organization that protects nesting sea turtles. They were delighted to find and report a new nest, and they later learned that it contained 87 eggs. In Melissa's case, this newly sparked interest led to her creation of a book about sea turtle tourism. Another acquaintance of mine, Patricia Laudano, became interested in the American Woodcock when one of the well-camouflaged ground dwellers startled her with its sudden flight up into her face. She has studied this unusual and compelling species for more than 15 years, and now this president of our local chapter of the National Audubon Society is called upon to apply her expertise to monitoring the birds' local population. She gives talks, too, the culmination being a chilly walk at dusk to see and hear the male American Woodcock's wooing ritual.

There are countless, unnamed citizen scientists—and surely only a minute percentage has books or lecture schedules underway—monitoring orchids, collecting soil, and performing flock head counts. It gives me great hope to think of all of these commitments, all of these hopeful actions, flowing in from all corners of the globe.

These points of data being collected about the natural world are an exponential leap beyond what George H. W. Bush once called "a thousand points of light" in praise of community action. One classic and time- tested example of citizen science at work is the Christmas Bird Count, sponsored by the Audubon Society and going strong since 1900. The yearly reports from birders have helped to inform the wildlife census, and in turn essential conservation efforts.

I have started with my own small steps in citizen science, driven by my growing love for this Earth. Gavin and

I reveled in The Great Backyard Bird Count last winter, stumbling around in two-foot deep snow with our binoculars and jotting down, with soggy mittens and numb fingers but also great joy, the species we found. In early spring I took up a much cushier assignment—I played part-time hooky while logged in to my day job from home, completely distracted by the titmice, woodpeckers, mourning doves, and kestrel that visited my yard as I made a list for Project FeederWatch. I have just been trained to monitor frogs in a local wetland this spring, but it's a bigger commitment. I had to register my wetland, take a written exam, and be able to identify 10 frog calls. (Much reviewing to do! I only just discovered that the deep, twangy, banjo-string call I often hear is a green frog, not a bullfrog.)

So, heroism can be writ large, but it can also be a reflection of many smaller actions that collectively have a significant impact. Perhaps if even the quietest appreciators of nature commit to attending and reporting with more vigor we can eventually call ourselves heroes of a sort, as our beloved Earth and its many fragile species inches slowly back from the brink of disaster. After all, as Christ said from the mountain, the meek shall inherit the earth.

A Peace on Hammocks

When I applied for my residency at Trail Wood, I included this sentence in the application: "One clever idea I'm determined to borrow from Teale's *A Naturalist Buys an Old Farm* is the observation of woodland companions from the vantage point of a hammock." I had some concern that the Connecticut Audubon Society might judge me lazy, but they gave me the benefit of the doubt. I moved into Trail Wood later that year.

On the day before I moved in, I acquired a cheap, rainbow-colored hammock reminiscent of ponchos brought home as souvenirs from Mexico. There was no chance I'd forget to pack it; it stood out like a party piñata among the other items awaiting the suitcase.

I read that Teale got his hammock at an army-navy store, so it was likely khaki in color or even camouflage, blending right in. I also had the impression that it was fairly deep, like a little cloth cave into which he could descend, sneaking peeks from time to time over the edge. Third, I surmised that the man knew how to tie a good knot.

If we were making comparisons here, I was already striking out. In case the animals hadn't already noticed the textile fiesta, I put on an extended display of what I now refer to as the Devil's Macramé, trying to fake knots that, owing to my lack of Girl Scout participation, I had never actually tied before. I was reminded of my extra poundage and the need to figure out a nonslipping knot when several times I descended to the ground with a thump. Increasingly giddy but also increasingly determined, I sat laughing and talking to myself on the ground after each fall. So much for my grand plan of clever, covert animal observation.

I finally surprised myself with good knots and tried to pretend I wasn't riding a rather shallow platform only two feet off the ground. The point was, I was in the hammock and it stayed put. Back on track for nature surveillance, I tried to make myself look small and unobtrusive. Almost immediately, and reminiscent of Teale's account, a wood-chuck stood up and spent a decided amount of time staring at me. It moved closer than it had when I wrote at the picnic table, probably figuring that a clearly challenged, awkward-ly suspended middle-aged woman wasn't likely to get up and give chase anytime soon.

I placed my binoculars and camera on my belly and looked around. How relaxing! Being suspended in a gen-tle sling felt cozily embryonic. Birds flew overhead, close to me as if on a dare. Robust breezes had the trees swaying gently, and I mused about how long it had been since I had simply listened to trees in the wind. Butterflies in particular seemed quite attracted to my recumbent form—maybe the hammock's vibrancy had some effect, but I also know that butterflies crave the salt on our skin. Whatever the enticement, I heard them fluttering below the canvas and a few joined me inside its fold. One large brown one spent quite a lot of time tickling my legs.

I wondered lazily whether birds take naps. I heard very few, compared with the busy choruses I'd witnessed at dawn and dusk. Suggestible to my own train of thought, I fell into a light, quick, midday sleep. When I awoke, five turkeys were silently marching in a row just past my feet. They disappeared around the bend.

The Trail Wood experience was a vast and rare gift that continues to make me smile; the hammock was the icing on this sweet cake. There was no one there to poke fun at my ineptitude, no one to consider me inefficient or lazy for choosing to ponder nature horizontally. There was no one to look at my pale, awkwardly placed legs or nerdy outdoor wardrobe and decide I didn't match the ideal picture of a nature lover or writer. There was no time limit, except the common-sense caution suggested by a forecasted storm. Job, family, bank account—all took a distant back seat to the moment at hand.

I have extracted some lessons from my brief ebullient time floating among the outdoor sights and sounds I cherish. The gist: even when life demands that you remain upright and moderately austere, take every opportunity to seize the peace of hammocks. Carry a hammock—or at least the peace it brings—wherever you go. Bring into the outdoors the spirit of the happy, if temporary, home for thinking and listening and looking and learning. All are practices that can sustain us in our daily lives and, in turn, ready us to help our welcoming but weary earth.

Acknowledgments

The following publications published original versions of the listed pieces:

"Close to the Edge" was published on *naturewriting.com*, January 2016
"Ants, Plants, and Pescetarians" was published in *Whole Life Times*, October/November 2014
"The Way in the Woods: A Week with Edwin" was published in *Connecticut Woodlands*, the magazine of the Connecticut Forest & Park Association, Winter 2016

These pieces had their origins from my writer-in-residence experience at Trail Wood, a Connecticut Audubon Property and former home of Edwin Way Teale, during the summer of 2015. The residency was sponsored by The Connecticut Audubon Society, and I remain eternally grateful for this chance to reconnect with nature, write, and live in the timeless presence of the Teales and their lifelong mission:

"A Curious Path"
"Beckoning the Mountain"
"The Way in the Woods: A Week with Edwin"
"A Peace on Hammocks"

Special thanks to *Connecticut Woodlands* editor and writing teacher Chris Woodside for illuminating the walking and writing connection!

Bibliography

Books & Print Periodicals

Rick Bass. *A Thousand Deer: Four Generations of Hunting and the Hill Country*. Austin, Texas: University of Texas Press, 2012.

Hal Borland. *Sundial of the Seasons*. Philadelphia: JB Lippincott Co, 1964.

Hal Borland. *Twelve Moons of the Year*. New York: Alfred A. Knopf, 1979.

C Byington. "First Flight for a New Birder." *Nature Conservancy*. November 2015.

Edwin H. Dodd, Jr. *Of Nature Time and Teale*. New York: Dodd, Mead & Company, 1966.

Jane Goodall. *Reason for Hope*. New York: Warner Books, 1999.

Ursula Goodenough. *The Sacred Depths of Nature*. New York: Oxford University Press, 1998.

Brenda Z. Guiberson. *Cactus Hotel*. New York: Square Fish, 1993.

Fred Hageneder. *The Living Wisdom of Trees*, London: Duncan Baird Publishers, 2005.

David George Haskell. *The Forest Unseen: A Year's Watch in Nature*. New York: Penguin Books, 2013.

Alexandra Horowitz. "A Dog's Nose View." In *On Looking: Eleven Walk with Expert Eyes*. New York: Scribner, 2013.

Barbara Hurd. *Stirring the Mud*. Athens, Georgia: University of Georgia Press, 2008.

Leo P. Kenney and Matthew R. Burne. *A Field Guide to the Animals of Vernal Pools*. Massachusetts: Massachusetts Division of Fisheries & Wildlife Natural Heritage & Endangered Species Program, Vernal Pool Association, 2009.

Richard Louv. *Last Child in the Woods*. Chapel Hill, NC: Algonquin Books, 2008.

Robert Macfarlane. *The Old Ways*. UK: Hamish Hamilton, 2012.

Bill Mollison. "Chapter 4: Pattern Understanding." In *Permaculture: A Designer's Manual*. Tyalgum, New South Wales, Australia: Tagari Publications, 1988.

Mary Oliver. *Thirst: Poems*. Boston: Beacon Press. 2006.

Charles Siebert. *Wickerby*. New York: Crown Publisher, 1997.

Shel Silverstein. *The Giving Tree*. New York: Harper Collins, 1999.

Edwin Way Teale. *A Walk Through the Year*. New York: Dodd, Mead, & Company, 1978.

Edwin Way Teale. *Days Without Time*. New York: Dodd, Mead & Company, 1948.

Larry Weber. *The Backyard Almanac: 365 Days of Northern Natural History*. Wrenshall, MN: Stone Ridge Press, 2014.

Marie Winn. *Central Park in the Dark: More Mysteries of Urban Wildlife*. New York: Farrar, Straus and Giroux, 2008.

Online References
(Unless otherwise noted, accessed August 2016)

"1966 Pulitzer Prizes," *Pulitzer.org*. http://www.pulitzer.org/ prize-winners-by-year/1966.

"22 Trees Growing Around Things," *Now That's Nifty*. http:// nowthatsnifty.blogspot.com/2010/02/22-trees-growing-around-objects.html.

"About Permaculture," *Permaculture.net*. http://www.permaculture.net/about/definitions.html.

Achenbach, Joel. "Why Do Many Reasonable People Doubt Science," *National Geographic*. March 2015. http://ngm. nationalgeographic.com/2015/03/science-doubters/ achenbach-text.

"Ants FAQ," *Uncle Milton*. http://unclemilton.com/support/ faq/ants/.

Applegate, Michael. "Anderson Hugs Trees, Takes to Skies for Fourth Slopestyle Gold," *The Aspen Times*. January 27, 2013. http://www.aspentimes.com/article/20130127/ SPORTS/130129901.

Attend definition. *Online Etymology Dictionary*. http:// etymonline.com/index.php?term=attend&allowed_in_ frame=0.

"A World Without Words," (interview with Jill Bolte Taylor), *RadioLab*. https://www.wnyc.org/radio/#/ ondemand/91729.

Barnes-Svarney, Patricia and Svarney, Thomas E. *The Handy Biology Answer* Book. https://www.papertrell.com/apps/ preview/The-Handy-Biology-Answer-Book/Handy%20 Answer%20book/.

Becker, Andrea. "Rates of Deforestation & Reforestation in the U.S.," *Seattle Post-Intelligencer*. http://education. seattlepi.com/rates-deforestation-reforestation-us-3804.html.

Bottini, Mike. "Notes on the Long Island Natural History Conference: Restoring the American Chestnut," *27East. com*. March 22, 2016. http://www.27east.com/news/ article.cfm/East-End/474015/Notes-On-The-Long-Island-Natural-History-Conference-Restoring-The-American-Chestnut.

Bound, Julian. "Photograph of the Day: Spiritual Roots," *National Geographic*. http://photography.national geographic.com/photography/photo-of-the-day/ buddha-statue-tree-thailand/.

Bramen, Lisa. 'What the Heck Was Manna, Anyway? The Unknown Fifth Question of the Passover Seder," *Smithsonian.com*. April 8, 2009. http://www.smithson ianmag.com/arts-culture/what-the-heck-was-manna-anyway-56294548/?no-ist.

Buffin, Aurélie. "The Secret Story of Ants: We Are Because We Share!," *derStandard.at*. http://derstandard. at/1296696461642/The-secret-story-of-ants.

Carpenter, Anita. "Coneucopia," *Wisconsin Natural Resources Magazine*. December 2007. http://dnr.wi.gov/ wnrmag/html/stories/2007/dec07/cones.htm.

Chamovitz, Daniel. "What a Plant Knows (and Other Things You Didn't Know About Plants) with Daniel C" (video). 2013. https://www.youtube.com/ watch?v=koeIq13aQxk.

"Christmas Tree Fact Sheets: Pine Spittlebugs," *The Ohio State University*. http://entomology.osu.edu/bugdoc/ Shetlar/factsheet/christmastree/pine_spittlebug.htm.

"Citizen Science," *National Geographic*. http://education. nationalgeographic.com/education/encyclopedia/ citizen-science/?ar_a=1.

"Common Snapping Turtle," *Connecticut DEEP*. http:// www.ct.gov/deep/cwp/view.asp?a=2723&q=469200.

Concord Grape History, *Concord Grape Association.* http://www.concordgrape.org/bodyhistory.html.

"Conifers for Wildlife" In Penn State Marcellus Shale Electronic Field Guide. http://www.marcellusfieldguide. org/index.php/guide/restoring_for_wildlife/conifers_ for_wildlife/.

Conniff, Richard. "The Wall of the Dead: A Memorial to Fallen Naturalists," *Strange Behaviors blog.* https:// strangebehaviors.wordpress.com/2011/01/14/the-wall-of-the-dead/.

Dante. Circle 9, cantos 31-34. *The University of Texas at Austin.* http://danteworlds.laits.utexas.edu/circle9.html.

"Eastern Newt," *The Animal Files.* http://www.theanimal files.com/amphibians/newts_salamanders/eastern_ newt.html.

"Eastern/Red Spotted Newt," *New Hampshire Fish and Game.* http://www.wildlife.state.nh.us/wildlife/profiles/ red-spotted-newt.html.

"Edwin Way Teale," *Find A Grave.* http://www.findagrave. com/cgi-bin/fg.cgi?page=gr&GRid=7320359.

"Facts about Rainforests," *The Nature Conservancy.* http:// www.nature.org/ourinitiatives/urgentissues/rainforests/ rainforests-facts.xml.

Feeley, Tivon. "Pine, Fir or Spruce Tree?," *Iowa University Extension News.* 2005. https://www.extension.iastate. edu/news/2005/nov/061401.htm.

"Fern Image Gallery,". http://www.home.aone.net. au/~byzantium/ferns/gallery/gallery3.html.

Field, Christopher B., Barros, Vincente R., Mach, Katherine J., Mastrandrea, Michael D. "Technical summary." *In: Climate Change 2014: Impacts, Adaptation, and Vulnerability.* http://www.ipcc.ch/pdf/assessment-report/ar5/wg2/WGIIAR5-TS_FINAL.pdf.

"Forest Bathing," *Healthy Parks Healthy People Central.*
http://www.hphpcentral.com/article/forest-bathing.

Gillis, Justin. "Restored Forests Breathe Life into Efforts
Against Climate Change," *New York Times.* Dec 23,
2014. http://www.nytimes.com/2014/12/24/science/
earth/restored-forests-are-making-inroads-against-
climate-change-.html?_r=1.

"Globally Almost 870 Million Chronically
Undernourished - New Hunger Report," *Global Policy
Forum.* 2012. https://www.globalpolicy.org/component/
content/article/217/51970-globally-almost-870-
million-chronically-undernourished-new-hunger-
report.html.

Hall, Whitney. "Planting Red Spruce at Cranesville
Preserve," *The Nature Conservancy.* http://www.nature.
org/ourinitiatives/regions/northamerica/unitedstates/
maryland_dc/explore/2015-red-spruce-planting-
passport-to-nature.xml.

Hardy, Thomas." Childhood Among the Ferns," *Poetry
Nook.* http://www.poetrynook.com/poem/childhood-
among-ferns.

Harton, Ron. "Hal Borland—Outdoor Writer."
Naturewriting.com. http://naturewriting.com/blog/
hal-borland-outdoor-writer/.

Hay, John. "Listening for Winter in the Silence of Birds;
Twelve Moons of the Year, By Hal Borland," *The
Christian Science Monitor.* 1980 . http://www.csmonitor.
com/1980/0114/011408.html.

Hu, Elise. "How the Sharing Economy Is Changing the
Places We Work," *NPR.* http://www.npr.org/blogs/
alltechconsidered/2013/11/14/244568645/how-the-
sharing-economy-is-changing-the-places-we-work.

Hughes, Ted "Fern," *Poet's House blog.* http://poetshouse.
blogspot.com/2006/03/ted-hughes-poems.html.

Iannotti, Marie. "Spittlebug," *Gardening on About.com*. http://gardening.about.com/od/insectpestid/a/Spittle bugs.htm.

"Introduction to Slugs and Snails," *University of Florida*. http://ipm.ncsu.edu/AG136/slugintr.html.

"Invasive Species," *United States Geological Survey*. http://www.nwrc.usgs.gov/topics/invasive_species/index.htm.

Jarvis, Alice-Azania. "Fern-mania," *Daily Mail*. http://www.dailymail.co.uk/news/article-2111859/Fern-mania-After-Victorians-said-cured-madness-boosted-love-life--explorers-risked-death-humble-fern-sexy-again.html#ixzz2Zj2EtrLi.

Kenlan, Peter. "A Field Guide to Conifers." http://www.bio.brandeis.edu/fieldbio/pkenlan/HTML/index.html.

Kubby, Steve. "Manna from Heaven," *Deoxy.org*. http://deoxy.org/manna.htm.

Kuo, Michael. "Phallus indusiatus," *MushroomExpert*. http://www.mushroomexpert.com/phallus_indusiatus.html.

Lawson, W. "Tracking Study Finds Ants Change Jobs as They Age," *MSN.com*. April 26, 2013, accessed October 7, 2013. http://news.msn.com/science-technology/tracking-study-finds-ants-change-jobs-as-they-age?stay=1.

Lehman, Eric D. "Leaves of Grass by Walt Whitman," *Umbrella*. 2008, Issue 6. http://www.umbrellajournal.com/spring2008/how_divine/EricD.Lehman.html.

Mayntz, Melissa. "Little Brown Job," *Birding at About.com*. http://birding.about.com/od/birdingglossary/g/lbj.htm.

Mayntz, Melissa. "Trees for Birds," *Birding at About.com*. http://birding.about.com/od/attractingbirds/a/Trees-For-Birds.htm.

"Native American Technology and Art. An Introduction to Tamarack Trees & Tradition," *Native Tech*. 2000. http://www.nativetech.org/willow/tamarack/tamarack.html.

"Northern White Cedar." http://www.bio.brandeis.edu/fieldbio/pkenlan/HTML/Cupressaceae/thuja_occidentalis.html.

"Obituary: Barbara Dodge Borland; Writer, 87," *New York Times*. http://www.nytimes.com/1991/02/14/obituaries/barbara-dodge-borland-writer-87.html.

"Oceans & Sea Level Rise," *Climate Institute*. http://www.climate.org/topics/sea-level/.

Pesaturo, Janet. "Woolly Oak Galls, Callirhytis lanata." *Our One Acre Farm*, October 2013. http://ouroneacrefarm.com/woolly-oak-galls-callirhytis-lanata/.

Pinopsida class, *Montana Field Guide*.. http://fieldguide.mt.gov/displayFamily.aspx?class=Pinopsida.

"Plant a Billion Trees," *The Nature Conservancy*. http://www.plantabillion.org/.

"Plants—How Could We Do Without Them?," *Botanic Gardens Conservation International*. January 2004. http://www.bgci.org/cultivate/article/390/.

"Purple Loosestrife," *National Wildlife Refuge Association*. http://refugeassociation.org/advocacy/refuge-issues/invasive-species/purple-loosestrife.

"Rana sylvatca Species Page," *The Virtual Nature Trail at Penn State*. 2002. http://www.psu.edu/dept/nkbiology/naturetrail/speciespages/woodfrog.htm.

"Red-Spotted Newt," *The Hiker's Notebook*. http://www.sierrapotomac.org/W_Needham/RedSpottedNewt_100705R.htm.

Riely, Elizabeth Gawthrop. "He Sowed, Others Reaped: Ephraim Bull's Concord Grapes," *Edible Boston*. http://www.edibleboston.com/edible-traditions-621/.

Ross, Michelle. "Ants and Their Dead," *A Moment of Science.* April 19, 2013. http://indianapublicmedia.org/amomentofscience/ants-and-their-dead/.

Roth, Sally. "All about Sparrows," *Birds and Blooms.* http://www.birdsandblooms.com/birding/bird-species/sparrows/sparrows/.

"Salad Burnet," *Cooks Info.* http://www.cooksinfo.com/salad-burnet.

"Salad Burnet History & Beverages," *Z Fooding.* December 9, 2013. http://zfooding.blogspot.com/2013/12/salad-burnet-history-beverages.html.

Schnall, Marianne. Exclusive Interview with Dr. Jane Goodall. *Huffington Post.* June 1, 2010. http://www.huffingtonpost.com/marianne-schnall/exclusive-interview-with_b_479894.html.

Schofield Edmund A. " 'He sowed, others reaped': Ephraim Wales Bull and the Origin of the 'Concord' Grape." http://arnoldia.arboretum.harvard.edu/pdf/articles/749.pdf.

Schwarzlose, Rebecca. "Can You Name That Scent?," *Gardens of the Mind blog.* January 27, 2014. http://gardenofthemind.com/2014/01/27/can-you-name-that-scent/.

"Slugs," *Royal Horticultural Society.* https://www.rhs.org.uk/advice/profile?pid=228.

"Sol LeWitt Designed Synagogue is Focus of World Premiere Documentary in Madison," *Connecticut by the Numbers.* June 10, 2015. http://ctbythenumbers.info/2015/06/10/sol-lewitt-designed-synagogue-is-focus-of-world-premiere-documentary-in-madison/.

Stafford, Tom. "Why Can Smells Unlock Forgotten Memories?," *BBC.com.* March 13, 2012. http://www.bbc.com/future/story/20120312-why-can-smells-unlock-memories.

"The Three Phases of the Vernal Pool Ecosystem," Sacramento Splash. http://www.sacsplash.org/post/three-phases-vernal-pool-ecosystem.

Thomas, Dylan. "Fern Hill," *Poets.org*. http://www.poets.org/viewmedia.php/prmMID/15378.

Thomas, Tracey. "Deep River's XYZ Bank Robber Rides Again," *Hartford Courant*. June 1, 1996. http://articles.courant.com/1996-06-01/news/9606010276_1_would-be-bank-robbers-warning-letters.

"Transcript for Ellen Davis and Wendell Berry: The Poetry of Creatures," November 24, 2011. http://www.onbeing.org/program/poetry-creatures/transcript/288. Interview by Krista Tippett.

"Transcript for Ellen Langer — Science of Mindlessness and Mindfulness," September 10, 2015. http://www.onbeing.org/program/ellen-langer-science-of-mindlessness-and-mindfulness/6332. Interview by Krista Tippett.

"Transcript for Indigo Girls: Music and Finding God in Church and Smoky Bars," October 3, 2013.http://www.onbeing.org/program/indigo-girls-music-and-finding-god-in-church-and-smoky-bars/transcript/8159#main_content. Interview by Krista Tippett.

"Transcript for Joanna Macy — A Wild Love for The World," November 6, 2014. http://www.onbeing.org/program/joanna-macy-a-wild-love-for-the-world/transcript/7022. Interview by Krista Tippett.

"Transcript for Mary Oliver: Listening to The World," February 5, 2015. http://www.onbeing.org/program/mary-oliver-listening-to-the-world/transcript/7271. Interview by Krista Tippett.

University of California Museum of Paleontology and the National Center for Science Education. "A Case Study of Coevolution: Squirrels, Birds, and the Pinecones They

Love (2 of 2)," *Understanding Evolution*. http://evolution.
berkeley.edu/evolibrary/article/evo_35.

"Vernal Pools," *EPA*. https://www.epa.gov/wetlands/vernal-
pools.

"Vernal Pool Wetlands," *REMA Ecological*, 2000. http://
www.remaecological.com/downloads/Vernalpools.pdf.

"Whirligig Beetles," *Missouri Department of Conservation*.
http://mdc.mo.gov/discover-nature/field-guide/whirli
gig-beetles.

"Wicked Big Puddles," *The Vernal Pool Association*. http://
vernalpool.org/vernal_1.htm.

Woodson, Dotty. "Ask an Expert: Response to 'Immature
Pine Cones Dropping from My Trees'," *Extension*. 2013.
https://ask.extension.org/questions/132269#.VZLJK
ZU5LIX.

"Why Humans Don't Smell as Well as Other Mammals:
No New Neurons in the Human Olfactory Bulb," *Science
Daily*. May 24, 2012. http://www.sciencedaily.com/
releases/2012/05/120524092222.htm.

Other Media

Lord of the Ants, directed by David Dugan. 2008. http://
www.pbs.org/wgbh/nova/nature/lord-ants.html.

Vegucated, directed by Marisa Miller Wolfson. 2010. http://
www.imdb.com/title/tt1814930/.

About the Author

Katherine Hauswirth's nature writing arises largely from long walks in Connecticut. Her work focuses on connection and contemplation inspired by the natural world. She has been published in *The Christian Science Monitor, The Day, Orion online, Whole Life Times, Connecticut Woodlands, Shoreline Times, Seasons,* and *The Wayfarer.* Her blog, *First Person Naturalist,* is a reflection on experiencing and learning about nature. Katherine's writing has been awarded with artist residencies at Trail Wood (Connecticut Audubon's Edwin Way Teale memorial sanctuary) and Acadia National Park in Maine. A native New Yorker, she moved to the Connecticut River Valley 20 years ago. She is increasingly enamored of her adopted hometown, Deep River, where she lives with her husband and son.

HOMEBOUND PUBLICATIONS

Ensuring that the mainstream isn't the only stream.

At Homebound Publications, we publish books written by independent voices for independent minds. Our books focus on a return to simplicity and balance, connection to the earth and each other, and the search for meaning and authenticity. Founded in 2011, Homebound Publications is one of the rising independent publishers in the country. Collectively through our imprints, we publish between fifteen to twenty offerings each year. Our authors have received dozens of awards, including: *Foreword Reviews'* Book of the Year, Nautilus Book Award, Benjamin Franklin Book Awards, and Saltire Literary Awards. Highly-respected among bookstores, readers and authors alike, Homebound Publications has a proven devotion to quality, originality and integrity.

We are a small press with big ideas. As an independent publisher we strive to ensure that the mainstream is not the only stream. It is our intention at Homebound Publications to preserve contemplative storytelling. We publish full-length introspective works of creative non-fiction as well as essay collections, travel writing, poetry, and novels. In all our titles, our intention is to introduce new perspectives that will directly aid humankind in the trials we face at present as a global village.

CPSIA information can be obtained
at www.ICGtesting.com
Printed in the USA
JSHW051449150522
25811JS00008B/155